157 Topics in Current Chemistry

W0051344

Chemical Applications of Nuclear Probes

Editor: K. Yoshihara

With contributions by
G. Harbottle, Y. Ito, J. I. Kim, R. Klenze,
R. Stumpe, N. Suzuki, K. Yoshihara

With 85 Figures and 23 Tables

Springer-Verlag Berlin Heidelberg GmbH

This series presents critical reviews of the present position and future trends in modern chemical research. It is addressed to all research and industrial chemists who wish to keep abreast of advances in their subject.

As a rule, contributions are specially commissioned. The editors and publishers will, however, always pleased to be receive suggestions and supplementary information. Papers are accepted for "Topics in Current Chemistry" in English.

ISBN 978-3-662-15051-1   ISBN 978-3-540-46987-2 (eBook)
DOI 10.1007/978-3-540-46987-2

This work is subject to copyright. All rights are reserved, whether the whole or part of the material is concerned, specifically the rights of translation, reprinting, re-use of illustrations, recitation, broadeasting, reproduction on microfilms or in other ways, and storage in data banks. Duplication of this publication or parts thereof is only permitted under the provisions of the German Copyright Law. of September 9, 1965, in its current version, and a copyright fee mus- always be paid

© Springer-Verlag Berlin Heidelberg 1990
Originally published by Springer-Verlag Berlin Heidelberg New York in 1990
Softcover reprint of the hardcover 1st edition 1990

The use of registered namens, trademarks, etc. in this publication does not imply, even in the absence of a specific statement, that such names are exempt from the relevant protective laws and regulations and therefore free for general use.

2151/3020-543210 — Printed on acid-free paper

Guest Editor

Prof. Dr. *Kenji Yoshihara*
Department of Chemistry, Faculty of Science,
Tohoku University, Sendai 980, Japan

Editorial Board

Prof. Dr. *Michael J. S. Dewar*	Department of Chemistry, The University of Texas Austin, TX 78712, USA
Prof. Dr. *Jack D. Dunitz*	Laboratorium für Organische Chemie der Eidgenössischen Hochschule Universitätsstraße 6/8, CH-8006 Zürich
Prof. Dr. *Klaus Hafner*	Institut für Organische Chemie der TH Petersenstraße 15, D-6100 Darmstadt
Prof. Dr. *Shô Itô*	Faculty of Pharmaceutical Sciences, Tokushima Bunri University, Tokushima 770, Japan
Prof. Dr. *Jean-Marie Lehn*	Institut de Chimie, Université de Strasbourg, 1, rue Blaise Pascal, B.P. Z296/R8, F-67008 Strasbourg-Cedex
Prof. Dr. *Kurt Niedenzu*	University of Kentucky, College of Arts and Sciences Department of Chemistry, Lexington, KY 40506, USA
Prof. Dr. *Kenneth N Raymond*	Department of Chemistry, University of California, Berkeley, California 94720, USA
Prof. Dr. *Charles W. Rees*	Hofmann Professor of Organic Chemistry, Department of Chemistry, Imperial College of Science and Technology, South Kensington, London SW7 2AY, England
Prof. Dr *Fritz Vögtle*	Institut für Organische Chemie und Biochemie der Universität, Gerhard-Domagk-Str. 1, D-5300 Bonn 1

Table of Contents

Chemical Nuclear Probes Using Photon Intensity Ratios

Kenji Yoshihara

Department of Chemistry, Faculty of Science, Tohoku University Sendai, 980 Japan

Table of Contents

Intensity ratios of photons emitted from radionuclides or from mesonic atoms are sometimes influenced by chemical environments. Similar phenomena are also observed in particle induced X-ray emission. The understanding of these experimental results has been developing recently, and some attempts to apply these phenomena for analytical purposes have appeared. Interesting facets of the chemical effects of the X-ray intensity ratios in the study of electronic K and L X-rays are touched on in this

Topics in Current Chemistry, Vol. 157
© Springer-Verlag Berlin Heidelberg 1990

Kenji Yoshihara

review in order to provide readers with the fundamental concepts of the phenomena. The chemical effects of electronic X-rays induced by particle bombardment which are topics of ion beam studies in material science are included. Chemical effects of muonic and pionic X-rays which are generated on mesonic orbital transitions are also described briefly. The sum peak method which is based on the principle of perturbed angular correlation and practicable with a simplified measurement assembly is applied to various chemical and biochemical substances. The photon intensity ratio in the sum peak method is studied both experimentally and theoretically. The present status of the study of the sum peak method is reviewed.

1 Introduction

Intensity ratios of photons emitted from some of radioactive nuclides are known to vary with environmental conditions under which the sources are placed. Such phenomena are interesting to fundamental researchers from the viewpoint of exploring the causes of their occurrence. In the cases of electron capture (EC) and isomeric transition (IT) decays, efforts have been devoted mainly to describing the determining factors for changes of photon intensity ratios (K_β/K_α, etc.) in detail.

However, the variation of the photon intensity ratios with the environment is also noteworthy for its possible application to analysis of chemical states. A few papers with such a practical purpose in mind have been published, although it remains to solve some problems for the method to become applicable for the purpose of reliable chemical speciation.

The objectives of photon intensity ratio change to be discussed in this chapter are not limited to those observed when EC or IT decay occurs. Electron, proton, or heavier ion bombardment on substances can produce X-rays and cause changes in their intensity ratio to some extent. These particle bombardment phenomena are included here as related to effects resulting from the influence of outer electrons on inner electron-nucleus interactions. (The X-ray excitation method is not described here, but is dealt with elsewhere).

The topics which are discussed in this chapter are concerned with energetic X-rays produced by negative muon or pion capture by nuclei; their energy is far beyond the X-ray region in an ordinary sense, lying in the γ-ray region. However, they are analogous to electronic X-rays with fine structures. Interesting features of muonic and pionic X-ray production have been revealed by recent studies of muonic and pionic atom chemistry. These X-rays are known as mesonic (mesic) X-rays and their chemical effects are described here briefly.

Another subject which will be of interest for those who wish to apply nuclear chemistry for analytical purposes, is the "sum peak method". The principle of this method is based on a perturbed angular correlation (PAC) for two γ-emissions in cascade decay from a radioactive nucleus. The emission angle between the two γ's has a distribution pattern which reflects the mode of radiative decay, as well as depending on the environmental conditions. The sum peak which is seen in a γ-ray spectrum as a result of simultaneous detection of the two γ-rays as one event, is therefore influenced by the environments in which the source is placed. In the sum peak method, intensity ratios of the sum peak to the single peak can be used and changes in the ratios due to the environments can be observed.

In summarizing the chapter topics, all share the following common features:

1) The observation objectives are related to the phenomena resulting from nucleus-electron interactions with emphasis on effects of outer electrons.
2) Chemical influences appear in the photon intensity ratios.
3) The effects are detectable using simple measurement systems such as photon spectrometers equipped with Si(Li), pure germanium, or Ge(Li) detectors.

2 Electronic X-ray Intensity Ratios

The X-ray intensity ratio K_β/K_α has been experimentally studied for various elements, and compared with theoretically obtained values. Good agreement has been observed between the experimental and theoretical ratios [1]. A general trend of data is shown in Fig. 1. While the ratio clearly increases with increasing atomic number Z, it has been recognized that the ratio depends on the mode of excitation of the atom concerned [2], and also on the chemical environment to some extent in certain d-block elements [3].

Prior to reviewing recent developments from studies in this research field, the reader's attention is directed to Table 1 which provides a convenient listing of the notation for

Table 1. Occurrence of electronic X-rays

Notation	Transition	
	Shell	Orbital
$K\alpha_1$	$L_3 \rightarrow K$	$2p_{3/2} \rightarrow 1s$
$K\alpha_2$	$L_2 \rightarrow K$	$2p_{1/2} \rightarrow 1s$
$K\beta_1 \longrightarrow K\beta_1'$	$M_3 \rightarrow K$	$3p_{3/2} \rightarrow 1s$
$K\beta_2$	$N_{2+3} \rightarrow K$	$4p_{1,3/2} \rightarrow 1s$
$K\beta_3$	$M_2 \rightarrow K$	$3p_{1/2} \rightarrow 1s$
$K\beta_4 \longrightarrow K\beta_2'$	$N_{4+5} \rightarrow K$	$4p_{3/2} \atop 4d_{3/2} \Big\} \rightarrow 1s$
$K\beta_5$	$M_{4+5} \rightarrow K$	$3d_{3,5/2} \rightarrow 1s$
$L\alpha_1$	$M_5 \rightarrow L_3$	$3d_{5/2} \rightarrow 2p_{3/2}$
$L\alpha_2$	$M_4 \rightarrow L_3$	$3d_{3/2} \rightarrow 2p_{3/2}$
$L\beta_1$	$M_5 \rightarrow L_2$	$3d_{5/2} \rightarrow 2p_{1/2}$
$L\beta_2$	$N_5 \rightarrow L_3$	$4d_{5/2} \rightarrow 2p_{3/2}$
$L\beta_3$	$M_4 \rightarrow L_1$	$3p_{3/2} \rightarrow 2s$
$L\beta_4$	$M_2 \rightarrow L_1$	$3p_{1/2} \rightarrow 2s$
$L\beta_6$	$N_1 \rightarrow L_3$	$4s \rightarrow 2p_{3/2}$
$L\gamma_1$	$N_4 \rightarrow L_2$	$4d_{3/2} \rightarrow 2p_{1/2}$
$L\gamma_2$	$N_2 \rightarrow L_1$	$4p_{1/2} \rightarrow 2s$
$L\gamma_3$	$N_4 \rightarrow L_1$	$4d_{3/2} \rightarrow 2s$
$L\gamma_4$	$O_{2,3} \rightarrow L_1$	$5p_{1,3/2} \rightarrow 2s$

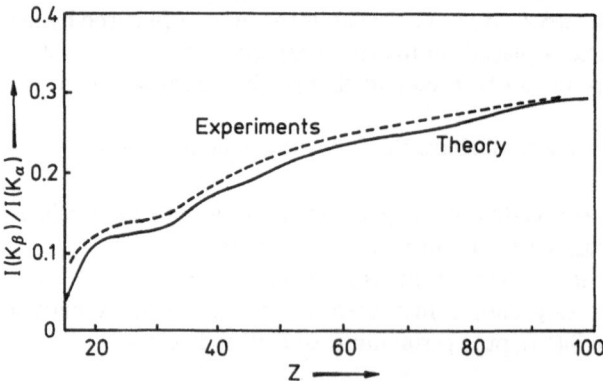

Fig. 1. Dependence of the X-ray intensity ratio K_β/K_α on atomic number Z^1

X-rays and electronic transitions. The generation of characteristic X-rays is ascribed to the production of an original electron vacancy in the inner electron shell, followed by filling of the vacancy with an outer electron. Multiple vacancy production and subsequent appearance of satellite lines are not included in the table, but they are treated later.

2.1 Electron Capture Decay

In EC decay, the chemical effects of the X-ray intensity ratio K_α/K_β was first observed by Tamaki et al. [4, 5] for various chromium compounds labeled with ^{51}Cr. A Si(Li) detector was used to measure the X-rays emitted from the nuclide. When the measured data of K_β/K_α are plotted against the formal oxidation number of chromium, it is seen that the ratio generally increases with the chromium valence number. This is shown in Fig. 2. A notable exception is, however, metallic chromium, where the formal oxidation number does not reflect the electronic state surrounding the chromium atom at all.

Lazzarini et al. [6] also studied the chemical effect on the X-ray intensity ratios in chromium compounds labeled with ^{51}Cr or in ^{51}Cr-doped materials. They pointed out that the intensity ratio might be influenced by factors such as crystal parameters of the matrix surrounding the decaying atom. They also performed similar experiments for ^{55}Fe-labeled compounds and for ^{55}Fe-doped crystals, and found chemical effects on the X-ray intensity ratio K_β/K_α in these cases as well.

As to manganese which is just between chromium and iron in the periodic table, it is not surprising that its K_β/K_α X-ray intensity ratio increases with increasing formal oxidation number; still, only a slight effect was found [7].

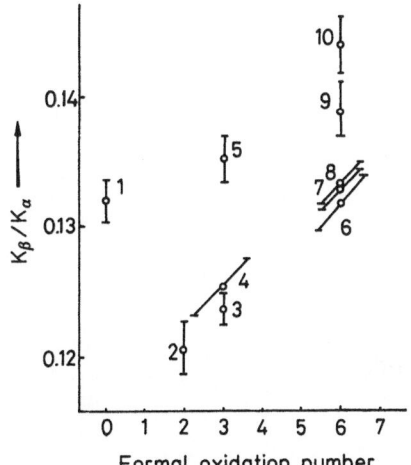

Formal oxidation number

Fig. 2. Chemical effect on the K_β/K_α X-ray intensity ratios in ^{51}Cr-labeled chromium compounds.
1. Cr metal, 2. $CrCl_2$, 3. $CrCl_3$, 4. $KCr(SO_4)_2$ · 12 H_2O, 5. Cr_2O_3, 6. $(NH_4)_2Cr_2O_7$, 7. $K_2Cr_2O_7$, 8. K_2CrO_4, 9. $PbCrO_4$, 10. CrO_3

2.2 Isomeric Transition Decay

In IT decay from a metastable state to its ground state of a nucleus, an internal conversion process often occurs and as a result, characteristic X-ray emissions are observed. The chemical effect of the X-ray intensity ratio was first studied by Yoshihara and coworkers [8, 9, 10] in the case of the ^{99m}Tc → ^{99}Tc isomeric transition. The nuclide ^{99m}Tc possesses an interesting property in that the nuclear decay rate is influenced by environmental conditions [11] or by chemical states [12, 13]. As technetium is one of the second-row transition element, characterized by 4 d valence electrons, it would be expected to show a smaller chemical effect of the K_β/K_α X-ray intensity ratio than chromium and manganese which are characterized by 3 d valence electrons. However, its K_β X-ray can be broken down into two components, $K_{\beta 1}'$ and $K_{\beta 2}'$ (see Table 1), when measured with a Si(Li) or pure germanium detector, and the latter component is found to be more sensitive to the intensity ratio change than the former. In order to detect the effect, three technetium nuclides ^{99m}Tc, ^{97m}Tc, and ^{95m}Tc were examined in various chemical forms. Their decay schemes are shown in Fig. 3. ^{99m}Tc undergoes a two-step IT decay; ^{97m}Tc, one-step IT decay; and ^{95m}Tc, EC decay. Tc X-rays are emitted from ^{99m}Tc and ^{97m}Tc, whereas mainly Mo X-rays are emitted from ^{95m}Tc. The intensity ratio $K_{\beta 2}'/K_\alpha$ is plotted against the valence charge difference C_{eff} between technetium and its partner atom. C_{eff} is often referred to as 'ionicity' and it is defined by:

$$C_{eff} = v_f\{1 - \exp[-(\chi_A - \chi_B)^2/4]\} \tag{1}$$

where v_f is the formal oxidation number; χ_A and χ_B are electronegativities of the atoms A and B, respectively. The results for the three types of decay are shown in Fig. 4. The ratio increases linearly with increasing C_{eff} for each transition, except for metallic technetium. The slope for ^{99m}Tc in which the two-step IT occurs is about twice as

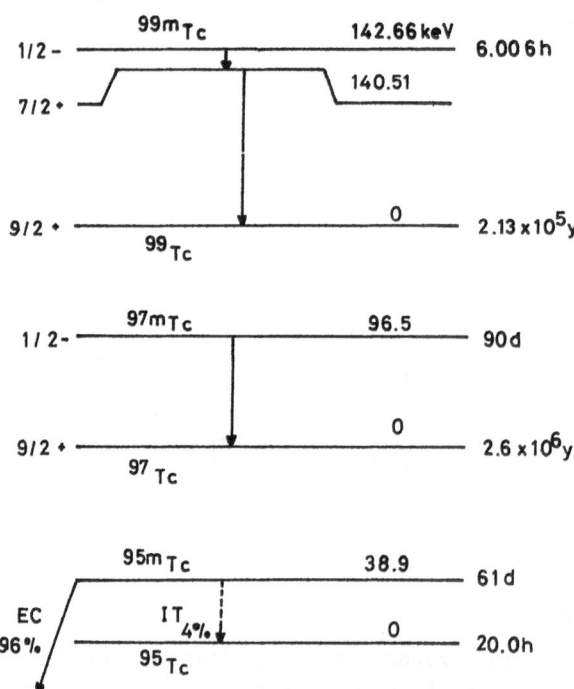

Fig. 3. Decay schemes for ^{99m}Tc, ^{97m}Tc, and ^{95m}Tc

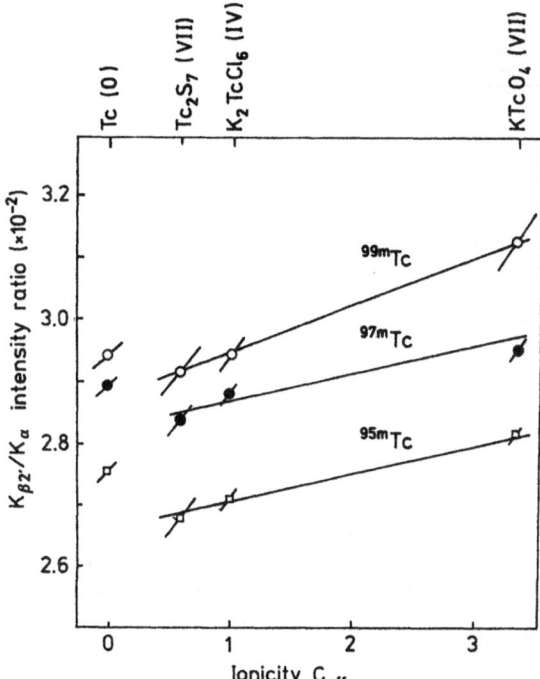

Fig. 4. Relation between the $K'_{\beta2}/K_\alpha$ X-ray intensity ratio and ionicity for ^{99m}Tc, ^{97m}Tc, and ^{95m}Tc

large as those for ^{97m}Tc and ^{95m}Tc which involve a one-step process. It is known that in successive IT processes, positive charge is stored on the atom, and this may cause the steeper slope.

Conversion electron spectroscopy of ^{99m}Tc in metallic technetium and ammonium pertechnetate revealed different patterns between them [14, 15]. Participation of O(2s) electrons was considered in the latter compound.

2.3 Electron, Proton, and Heavy Ion Bombardments

Inner electron shell vacancies can be produced by electron, proton, and heavy ion bombardments. When the vacancies are filled with electrons, characteristic X-ray emissions occur. Kiss et al. [16] observed the K_β/K_α X-ray intensity change due to chemical environments by bombarding various targets of titanium, chromium, and manganese with electrons. The magnitude of the chemical effect is about 6% for titanium ($TiO_2 - Ti$), for measurements performed using a Si(Li) detector.

Recently, Izawa et al. [17] studied the chemical effect of the $L_{\gamma1}/L_{\beta1}$ X-ray intensity ratio of molybdenum compounds by electron and proton bombardment. The X-ray spectra were measured with a curved crystal spectrometer (Johansson type, ADP crystal). $L_{\gamma1}$ and $L_{\beta1}$ are representative of variable and invariable peaks, respectively. $L_{\gamma1}$ arises from the transition $4d_{3/2} \to 2p_{1/2}$, in which 4d's are valence electrons. In Fig. 5 the $L_{\gamma1}/L_{\beta1}$ X-ray intensity ratio is plotted against ionicity for various molybdenum compounds. For electron bombardment, it is interesting to note that the ratio decreases with increase of ionicity. A similar tendency is also seen for proton bombardment, although the data are limited at present. This tendency may be related to a

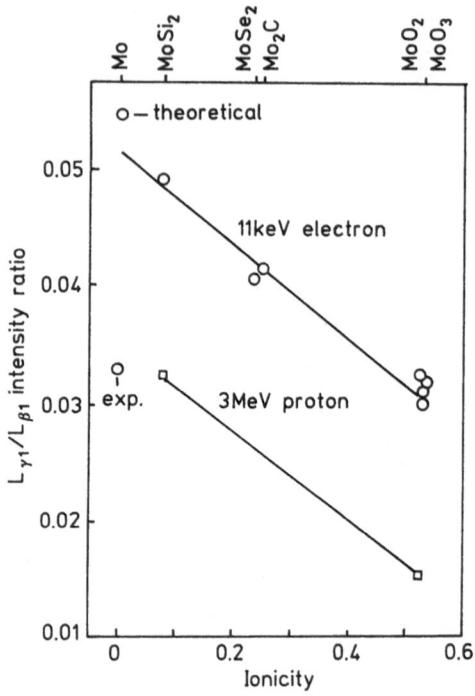

Fig. 5. Relation between the $L_{\gamma 1}/L_{\beta 1}$ X-ray intensity ratio and ionicity for electron- and proton-bombarded molybdenum compounds

direct transition of valence electrons. Again an exception in the trend is seen for metallic molybdenum, which exhibits a much lower value than that theoretically expected. As shown in Fig. 5, the intensity ratio changes significantly from 4.9×10^{-2} for $MoSi_2$ to 3.3×10^{-2} for MoO_3.

Similar experiments performed on niobium compounds by electron bombardment showed analogous trends [18] to those of molybdenum compounds.

Energy dependence of the K_β/K_α intensity ratio of SiO_2 for proton impact was studied by Benka [19] in the proton energy region 300–800 keV. The ratio increases with increase of proton energy by several percent at a maximum. No energy dependence was found for elemental silicon.

An important factor influencing the X-ray intensity ratio is the appearance of satellite lines which are not shown in the X-ray diagram (Table 1). Formation of double, triple, or multiple vacancies in the inner shell, or sometimes capture of electrons of counteratoms can produce these satellite lines. Other origins of the lines are discussed using molecular orbital considerations. Double K shell vacancy creation is observed in EC decay with a low probability ($P_{kk} \sim 10^{-4}$). This process is due to shake-off (SO) or to shake-up (SU) of the second K electron. More intense satellite lines are observed in ion bombardment.

When beryllium is bombarded with protons, electron vacancies of different types are created and filled with electrons of outer shells. For metallic beryllium and beryllium oxide, generated X-rays differ somewhat in energy and in satellite line formation. The results obtained by Ozawa et al. [20] for these targets are shown in Fig. 6. Satellite lines are clearly observed in the BeO spectrum, whereas the Be spectrum is rather

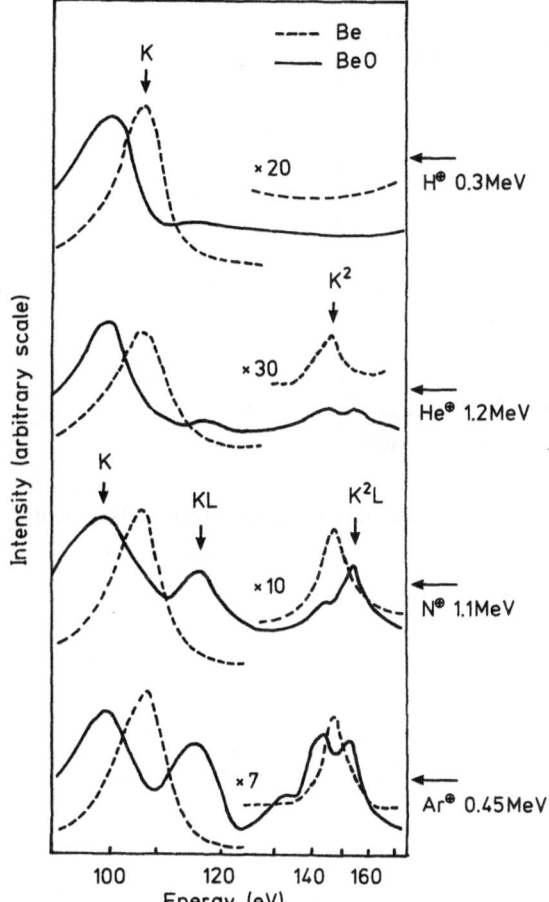

Fig. 6. Beryllium K X-ray spectra for proton, helium, nitrogen, and argon ion bombardment [20]

simple. Assignment of the peaks and energy shifts are shown in Table 2. Electrons from the L-shell of oxygen participate in the satellite formation.

Ozawa et al. [21] also performed similar experiments on boron compounds by bombarding with nitrogen ions (N^+). The measured spectrum is shown in Fig. 7. Satellite lines such as KL and K^2L appear in the case of boron nitride, while they do not appear in the case of elementary boron. The shoulder marked by K' is ascribed to transition of an electron of N(2s) to a vacancy of B(1s).

Table 2. Assignment of X-rays in Be and BeO

Initial vacancy	Transition	X-ray energy (eV)		Energy difference (eV)
		Be	BeO	
K	$Be(L) \rightarrow Be(K^{-1})$	108.5	104.4	—4.1
KL	$O(L) \rightarrow Be(K^{-1}L^{-1})$		116.0	
K^2	$Be(L) \rightarrow Be(K^{-2})$	146.1	142.3	—3.8
K^2L	$O(L) \rightarrow Be(K^{-2}L^{-1})$		153.8	

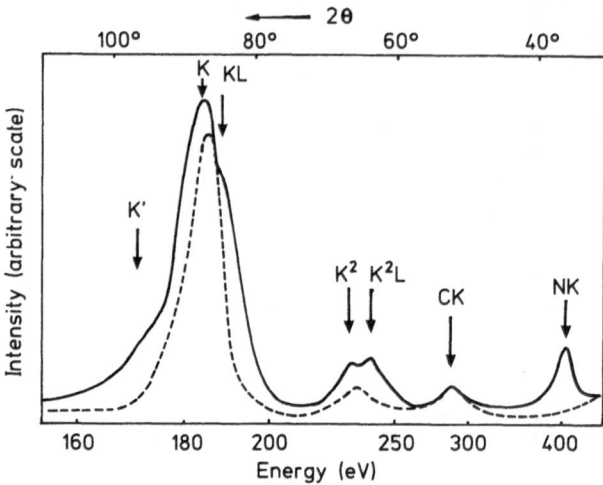

Fig. 7. K X-ray spectra [21] for elementary boron (dashed line) and boron nitride (solid line) bombarded with 0.7 MeV N$^+$

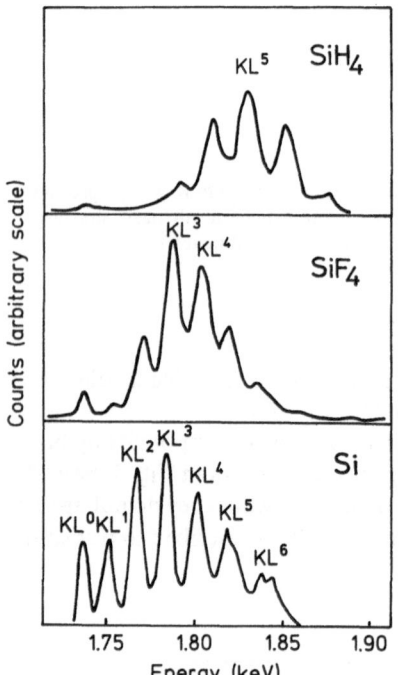

Fig. 8. Silicone K X-ray spectra with satellite lines in Cl$^+$-bombarded SiH$_4$ (gas), SiF$_4$ (gas), and Si (solid) [22]

The formation of multiple vacancies is an interesting process which has been studied intensively. Usually high-energy heavy ion bombardment on substances creates multiple vacancies in inner-sphere electron shells. Figure 8 shows the results on high-energy (53.4 MeV) Cl$^+$ bombardment of SiH$_4$ (gas), SiF$_4$ (gas), and Si (solid) [22]. The intensities of Si K$_\alpha$ lines depend on chemical environments. Theoretical interpretation of these phenomena will be given in Sect. 2.4.

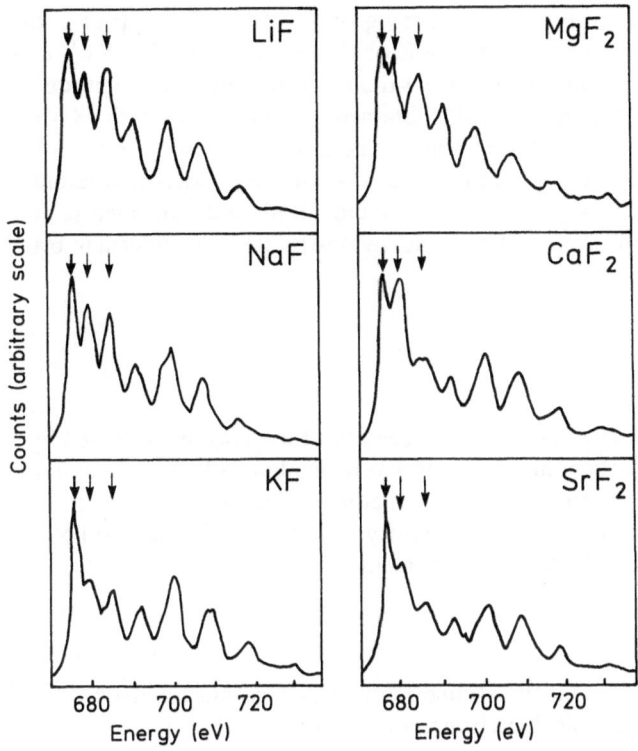

Fig. 9. Comparison of fluorine K_α X-ray spectra excited by 80 MeV Ar ions in solid targets of alkali metal and alkaline earth fluorides [23]. Arrows indicate KL^0, KL^1, and KL^2 lines

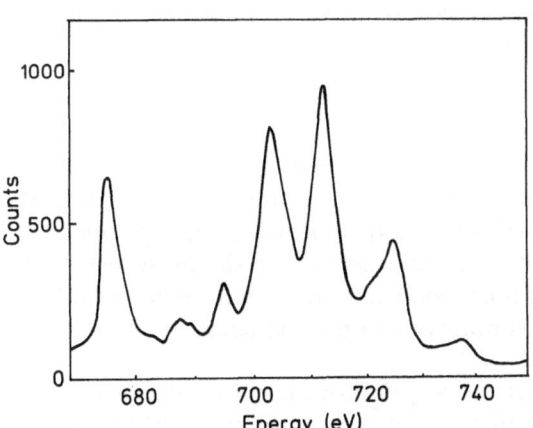

Fig. 10. Spectrum of fluorine K_α satellites excited by 48 MeV Mg ions in a SiF_4 (gas) target [23]

    Benka et al. [23] studied K X-ray spectra of solid and gaseous fluorine compounds by bombarding with 80 MeV Ar^{5+} ions and with 48 MeV Mg^{3+} ions. The K_α satellite and hypersatellite spectra were measured with a curved crystal spectrometer. The spectra obtained by bombarding with 80 MeV Ar^{5+} for various alkali metal and alkaline earth fluorides are shown in Fig. 9. The X-ray groups KL^0 to KL^7 appear in the order of increasing energy. Chemical influence is obviously seen for KL^0, KL^1, and KL^2, whereas it is not so prominent for the higher satellite groups. Figure 10 reproduces a spectrum of the K_α satellites in gaseous SiF_4 obtained by bombarding with 48 MeV Mg^{3+}. It is quite different from spectra for solids. The cause of enhanced KL^0, KL^1, and KL^2 peaks for the solid as compared to the gas is not fully understood. Similar chemical effects on satellite lines were observed by Uda et al. and attributed to two causes: change in ionization cross sections and reconstruction of molecular orbitals [23a].

2.4 Theoretical Studies

The chemical effect of the X-ray intensity ratio has been theoretically estimated by a number of researchers. The phenomenon itself can be explained by a rather simple treatment, and thereafter, by more sophisticated methods. The following equation [24] was proposed to estimate the K_β/K_α X-ray intensity ratio for particle-induced X-ray emission or photoionization in 3d transition elements:

$$I_{K\beta}/I_{K\alpha} = f_g(Z) \cdot g_{e,z} \qquad (2)$$

where $f_g(Z)$ is a term dependent on the atomic number (a constant for fixed Z) and $g_{e,z}$ is a term influenced by chemical environments as expressed by:

$$g_{e,z} = 1 + S_z \cdot K_d \cdot C_{eff} \qquad (3)$$

where S_z is a term due to contraction of the 3p orbitals by 3d charge delocalization. K_d is the 3d share of C_{eff}, the valence charge difference which can be obtained by Pauling's electronegatively conception:

$$C_{eff} = v_f\{1 - \exp[-(\chi_A - \chi_B)^2/4]\} \qquad (1)$$

where the notations v_f, χ_A, and χ_B were already explained in Sect. 2.2.

By applying these relationships to chromium, manganese, iron, and copper, Brunner et al. [24] obtained the values listed in Table 3. Experimental results by both PIXE (proton induced X-ray emission) and X-ray fluorescence are shown therein. If the value of $K_d = 0.52$ is used for calculations, good agreement between theoretical and experimental results is achieved for chromium and iron, although it is poorer for manganese and copper.

Mukoyama et al. [25] calculated the relative $K_{\beta2}/K_\alpha$ X-ray intensity ratios of technetium and molybdenum using the simplified model of Brunner et al. [24]. In this case the sensitivity S_z in Eq. (3) should be calculated for 4p contraction by 4d charge delocalization. The results for technetium and molybdenum thus calculated are shown in Table 4 together with experimental data obtained by Yamoto et al. [10]. The calculation qualitatively explains the chemical effect on $K_{\beta2}'/K_\alpha$ ratios experimentally obtained, except for the case of metallic technetium. As for the effect of differences in the excitation modes (IT and EC decays, etc.) Arndt et al. [2] gave the following explana-

Table 3. Relative $K\beta/K\alpha$ X-ray intensity ratios

Element	Constitutions	Experimental		Calculated ($K_d = 0.52$)
		PIXE	Fluorescence	
Cr	$Cr/K_2Cr_2O_7$	0.977	0.969	0.976
Mn	$MnS/KMnO_4$	0.950	0.953	0.974
	$MnSO_4/KMnO_4$	0.965	0.965	0.983
	$MnSO_4/MnSO_4$[a]	0.997	—	1
Fe	$Fe/FeSO_4$	0.987	0.986	0.984
	$FeSO_4/Fe_2(SO_4)_3$	0.996	—	0.992
	$Fe_2(SO_4)_3/Fe_2(SO_4)_3$[a]	0.993	—	1
Cu	Cu/CuS	0.973	—	0.997

[a] Different preparations

Table 4. Comparison of the calculated $K\beta_2/K\alpha$ ratios with the experimental values for Tc

Compound	Experimental		Calculated
	^{99m}Tc	^{97m}Tc	
Tc metal$/KTcO_4$	0.945 ± 0.017	0.977 ± 0.007	0.9605
$K_2TcCl_6/KTcO_4$	0.943 ± 0.017	0.977 ± 0.008	0.9730
$Tc_2S_7/KTcO_4$	0.940 ± 0.022	0.960 ± 0.010	0.9677

tion. Mo K X-rays are emitted following K-electron capture decay. The electronic configuration of Mo atoms differs from that produced by internal conversion or photo-ionization. The former is neutral, while the latter is a positive ion. Moreover, the probability of electron shake-off or shake-up differs for EC decay and for IT decay accompanied by internal conversion. The change in nuclear charge partially compensates for the change in the central potential due to formation of an inner electron shell vacancy.

A more sophisticated solution to this problem has been given by Band et al. [26] and by Mukoyama et al. [27] using the scattered-wave (SW) X_α molecular orbital method [28], and the discrete-variational (DV) X_α cluster method [29], respectively.

Table 5 shows the theoretical values calculated by Mukoyama et al. [27] for chromium compounds whose valences are II to VI. The value for each MO level is summed up to obtain the overall K_β/K_α ratio. Theoretical values are compared with experimentally obtained values. Except for CrO_3, the calculated results for chromium are qualitatively in agreement with the experimental data. The agreement is poorer for manganese than for chromium. The relative K_β/K_α intensity ratios obtained by Band et al. [26] are compared with those by Mukoyama et al. [27] and with experimental ones [4, 5, 7, 27] in Table 6. The theoretical values found by Band et al. are somewhat larger than those obtained by Mukoyama's group, and nearer to the experimental values, even though the former group used a simpler model than the latter. The reason of the discrepancy

13

Kenji Yoshihara

Table 5. $K\beta$ X-ray intensities in Cr (per $K\alpha$ intensity) and $K\beta/K\alpha$ ratios

T_d				O_h					
Level	K_2CrO_4	CrO_3	$K_2Cr_2O_7$	Level	CrO_2	Cr_2O_3	Level	$CrCl_2$	$CrCl_3$
$3t_2$	0.1141	0.1155	0.1180	$3t_{1u}$	0.1153	0.1159	$6t_{1u}$	0.1141	0.1147
$4t_2$	0.0041	0.0020	0.0032	$4t_{1u}$	0.0003	0.0002	$7t_{1u}$		0.0001
$5t_2$	0.0001	0.0003	0.0001	$5t_{1u}$	0.0008	0.0007	$8t_{1u}$	0.0005	0.0008
$6t_2$	0.0044	0.0025	0.0039	$6t_{1u}$			$9t_{1u}$		
$K\beta/K\alpha$	0.1227	0.1203	0.1252	K_β/K_α	0.1164	0.1168	K_β/K_α	0.1146	0.1156
expt.	0.1215	0.1264	0.1263	expt.		0.1185	expt.		0.1208

Table 6. Relative $K\beta/K\alpha$ intensity ratios

Compound	Theory		Experimental	
	Mukoyama et al.	Band et al.	Mukoyama et al. PIXE	Tamaki et al.
Cr_2O_3/K_2CrO_4	0.952		0.975	0.978 ± 0.020
CrO_3/K_2CrO_4	0.980		1.040	1.029 ± 0.021
$K_2Cr_2O_7/K_2CrO_4$	1.020		1.039	1.022 ± 0.021
$CrCl_3 \cdot 6\,H_2O/K_2CrO_4$	0.942	0.964	0.994	0.978 ± 0.020
CrO_2/K_2CrO_4	0.949			
$CrCl_2/K_2CrO_4$	0.934			
$MnCl_2 \cdot 4\,H_2O/KMnO_4$	0.937			0.971 ± 0.020
$MnO_2/KMnO_4$	0.960		0.973	1.007 ± 0.021
$K_2MnO_4/KMnO_4$	0.978			0.933 ± 0.021
$MnO/KMnO_4$	0.944			
$MnS/KMnO_4$	0.935	0.966		0.950 ± 0.007^a
				0.953 ± 0.003^b

[a] PIXE, by Brunner et al. [24]; [b] Fluorescence, by Brunner et al. [24]

is not fully understood. It seems that more detailed description of the model is needed in future studies.

Hartmann et al. [30] were the first to apply the X_α-SW calculation to estimate the chemically induced decay constant variation of ^{99m}Tc. The method meets the conditions to calculate the X-ray intensity ratio variation due to chemical environments as described above. Following their success in applying the X_α-SW method to inner shell problems in transition elements, a considerable volume of work has been published to explain such phenomena by MO calculations.

Hartmann et al. [31] theoretically studied the inner-shell ionization of KL^n type in silicon by the impact of heavy charged particles. As a result of the SCF-X_α-SW calculations, they obtained the valence electron configurations $a_1^2 t_2^6$ for SiH_4, $(1a_1)^2(1t_2)^6 \times (2a_1)^2(2t_2)^6 e^4 (3t_2)^6 t_1^6$ for SiF_4 and $(1a_1)^2(1t_2)^6(2a_1)^2(2t_2)^6 e^4$ for the silicon cluster. They calculated the total charge fractions in the central silicon sphere (Q_{si}), in the outer sphere (Q_{out}), in the ligand sphere (Q_L), and in the intersphere region (Q_{IS}). The results are shown in Table 7. q_v^s, q_v^p, and q_v^d denote the s-, p-, and d-type valence charge frac-

Table 7. Calculated totàl charges in the central silicon sphere (Q_{si}), in the outer (Q_{out}) and in the ligand sphere (Q_L), and in the intersphere region (Q_{IS})

		Core valency configuration								
		K^0L^0	KL^0	KL^1	KL^2	KL^3	KL^4	KL^5	KL^6	KL^7
Si	Q_{si}	12.85	13.01	13.12	13.14	13.06	13.09	13.28	13.43	13.48
cluster	q_v^s	1.23	1.43	1.56	1.68	1.77	1.72	1.86	1.88	1.99
	q_v^p	1.51	2.47	3.42	4.15	4.86	5.20	5.37	5.48	5.58
	q_v^d	0.09	0.11	0.14	0.21	0.42	1.06	2.05	3.06	4.00
	Q_L	12.11	12.05	11.98	11.92	11.86	11.77	11.67	11.55	11.43
	Q_{IS}	7.55	6.90	6.23	5.63	5.07	4.49	3.79	3.20	2.65
	Q_{out}	1.13	0.90	0.71	0.55	0.43	0.32	0.25	0.19	0.15
SiF_4	Q_{si}	12.18	11.98	11.80	11.80	11.83	11.88	11.99	12.10	12.21
	q_v^s	0.66	0.91	1.14	1.32	1.47	1.59	1.66	1.72	1.76
	q_v^p	0.96	1.46	1.67	2.57	3.27	3.90	4.41	4.76	5.01
	q_v^d	0.47	0.55	0.65	0.81	0.99	1.31	1.83	2.54	3.37
	Q_L	6.36	6.32	6.25	6.19	6.11	6.01	5.90	5.81	5.69
	Q_{IS}	10.42	10.07	9.82	9.23	8.70	8.26	7.67	7.05	6.47
	Q_{out}	1.97	1.69	1.38	1.19	1.01	0.83	0.72	0.63	0.55
SiH_4	Q_{si}	12.63	12.63	12.61	12.53	12.35	11.98	11.40	10.65	9.80
	q_v^s	0.93	1.18	1.39	1.56	1.69	1.80	1.87	1.92	1.95
	q_v^p	1.65	2.41	3.20	3.96	4.66	5.18	5.52	5.73	5.85
	q_v^d	0.05	0.04	0.03	0.02	0.01	—	—	—	—
	Q_L	0.12	0.10	0.08	0.05	0.04	0.02	0.01	0.004	0.002
	Q_{IS}	3.23	2.93	2.48	1.93	1.38	0.89	0.55	0.33	0.19
	Q_{out}	1.67	1.04	0.61	0.33	0.15	0.06	0.02	0.007	0.002

q_v^s, q_v^p, and q_v^d denote the s-, p-, and d-type valence charge fractions within the central silicon sphere

tions within the central silicon sphere. Total charge fractions in the table reveal that charge compensation of core-electron removal occurs by a valence charge transfer from the surroundings to the atom. The intersphere and outer sphere regions are the predominant contributors to this charge flow, indicating an adaptation of the valence electron cloud to the anisotropy of the surroundings. This makes the valence electrons more sensitive to the anisotropy of the surroundings, corroborated by the great amount of d-type valence charge (q_v^d). In SiH_4, however, the supply of ligand valence electrons is limited (c.f. the small value of q_v^d). Thus, the chemical environments are verified to have a great effect on the cascading deexcitation of highly excited atoms with multiple vacancies.

2.5 Applications to Chemical State Analysis

There are a few examples in which these phenomena have been applied to chemical state analysis. Collins et al. [32] measured the K_β/K_α X-ray intensity ratio for chromium using a thin film technique in order to correct accurately for self-absorption loss. They postulated that a good correlation could be obtained between the ratio and the valence of the element by this technique. Based on these findings, they further applied this 'living source' method to solve problems in hot atom chemistry. The method has an advantage that it can be performed *in situ* for unstable chemical species appearing in hot atom processes, without resorting to a dissolution procedure.

Collins et al. [33] also studied annealing processes in ^{51}Cr-doped crystals in which they doped ^{51}Cr(III) in/on K_2CrO_4 crystals and annealed them at elevated temperature. The doped ^{51}Cr(III) is finally found to be converted into ^{51}Cr(VI)-chromate. The intermediate processes are proposed as:

$$^{51}\text{Cr(III)-monomer} \rightarrow {}^{51}\text{Cr(III)-dimer}$$
$$\rightarrow \text{pseudo-}{}^{51}\text{Cr(VI)} \rightarrow {}^{51}\text{Cr(VI)}$$

based on inspection of the K_β/K_α intensity ratio change. In the above scheme, the 'pseudo-^{51}Cr(VI)' behaves as if it is in a Cr(III)-state in the solid for 'in situ' X-ray analysis, but it behaves as Cr(VI) on dissolution followed by radiochemical separation. The measured ratio of the K_β/K_α X-ray intensity in Fig. 11 differs from the results of wet chemical analysis in the region in which the pseudo-^{51}Cr(VI) is important. Thus, their observation on ^{51}Cr-doping in K_2CrO_4 crystals first point out the usefulness of the X-ray intensity ratio method. Care should be taken to apply this method to other systems, because we have a good reason to believe that the ionicity relation (see Sect. 2.4) rather than formal oxidation number plays a significant role in determining the ratio.

Tamaki et al. [34] also examined the 'in situ' method in thermal annealing of neutron-irradiated potassium chromate and of ^{51}Cr-doped potassium chromate. Variation of the X-ray intensity ratio in neutron-irradiated potassium chromate is not so prominent as in the ^{51}Cr-doped crystals from Collins' experiments, making the chemical effect somewhat obscured.

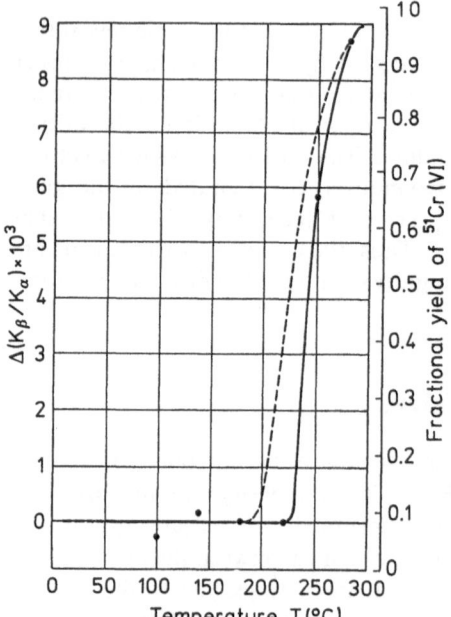

Fig. 11. Comparison of fractions of Cr(VI) obtained by X-ray 'in situ' analysis and by solution analysis in ^{51}Cr(III)-doped K_2CrO_4 after thermal annealing. The bold line is for 'in situ' analysis and the broken line for solution analysis [33]

3 Mesonic X-rays

Negatively charged muons and pions are captured by an atomic nucleus, resulting in formation of muonic and pionic atoms, respectively. During the capture process, muonic and pionic X-rays are generated by transitions from outer to inner orbitals of these mesonic atoms. (Sometimes these are referred to as 'exotic atoms'). The magnitude of the mesonic X-ray energy is 200–300 times as large as electronic X-ray energy, based on the mass ratio of the particle to an electron. In fact, the mesonic X-rays are in the energy region of γ-rays and are detected by ordinary γ-ray spectrometers equipped with Ge(Li) or pure germanium detectors.

Mesonic atoms can be regarded as new types of nuclear probes in material science dealing with chemical aspects. In this section, capture ratio problems in binary systems as well as X-ray intensity patterns in mesonic transitions are briefly described.

Figure 12 shows a schematic model of muonic and pionic atoms. Their orbitals are much smaller than those of electrons. Characteristic transitions between various levels are shown in Fig. 13. Due to the strong interaction of a pion with a nucleus, Lyman series pionic X-rays (n → 1) cannot be observed.

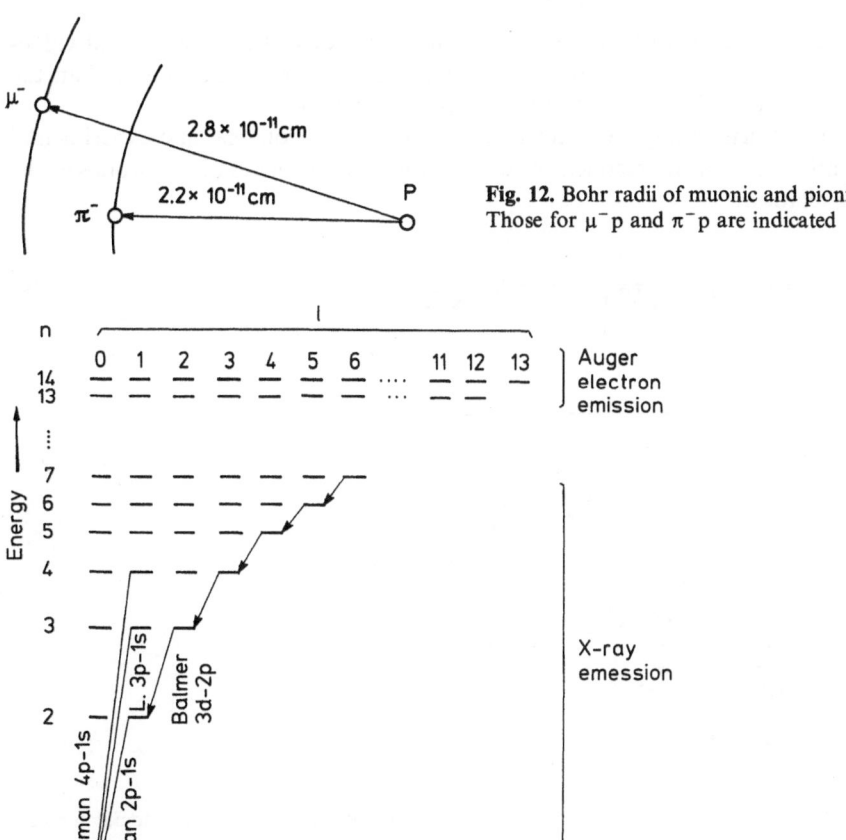

Fig. 12. Bohr radii of muonic and pionic atoms. Those for $\mu^- p$ and $\pi^- p$ are indicated

Fig. 13. Energy levels of a muonic atom and generation of muonic X rays

17

3.1 Capture Ratio

For the capture of a muon or pion by the atoms with Z_1 and Z_2 in a binary system, the Coulomb capture or atomic capture ratio is an important parameter. The capture ratio $A(Z_1, Z_2)$ is defined by:

$$A(Z_1, Z_2) = W_M(Z_1)/W_M(Z_2) \qquad (4)$$

where $W_M(Z_k)$ is the probability of formation of a mesonic atom $Z_k M^-$ when an M^- particle is stopped in a system consisting of elements Z_1 and Z_2 (Z_k denote both the element and its atomic number).

The capture ratio of the muon has been measured for chlorides as shown in Fig. 14. Fermi and Teller [35] assumed that the capture probability is proportional to the energy loss of the muon (or pion) near the corresponding atomic species. This leads to a relation between the capture ratio and Z_k as follows:

$$A(Z_1, Z_2) = Z_1/Z_2 \qquad (5)$$

However, this relation (often called the Z-law) does not fit the experimental values (dashed lines). The major disagreement is that the data are usually lower than the Fermi-Teller line for large Z and they change periodically.

Therefore, efforts to improve agreement between the theoretical and experimental capture ratios have been extended by several authors. Daniel et al. [36] proposed the following relation:

$$A(Z_1, Z_2) = \frac{Z_1^{1/3} \ln (0.57Z_1) \cdot R(Z_2)}{Z_2^{1/3} \ln (0.57Z_2) \cdot R(Z_1)} \qquad (6)$$

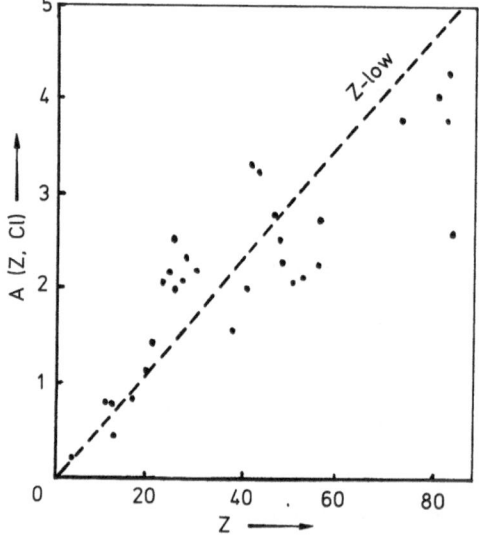

Fig. 14. Capture ratio and atomic number. The Fermi-Teller's Z-law is indicated by a broken line

where $Z_k^{1/3}$ is proportional to the cross-section of the atom when the model of Fermi and Thomas on the size of an atom ($\pi r^2 \propto Z_k^{1/3}$) is adopted. The logarithmic term relates to screening effects by the electron cloud. $1/R(Z_k)$ is an atomic radius term which influences capture of a muon through repulsion by a negatively charged electron. Equation (6) can explain the experimental data (dashed line in Fig. 15) relatively well, although it does not fit in the high Z region.

The dotted curve in Fig. 15 shows the calculated results of Schnewly et al. [37] who treated the effects of electrons using two terms: the first one relates to the valence electrons and the second, to the other electrons. In order to calculate the former term, they used ionicity σ estimated by $\sigma = 1 - \exp\{(\chi_1 - \chi_2)^2/4\}$, where χ_i ($i = 1, 2$) is the electronegativity. (Note that the 'ionicity' σ does not contain the formal exidation number v_f in Eq. (1).) Their calculation explains the periodical character of the capture ratio, although agreement is not complete in some places.

Further improvement of the theoretical calculation has been performed by von Egidy et al. [38] using a quantum mechanical approach. They assumed that electrons with binding energies less than 80 eV contribute with a weight which is a function of the electron binding energy, the electron quantum numbers n and l, and Z. Good agreement with experimental data is obtained as shown in Fig. 15 (solid line).

Imanishi et al. [39] pointed out that the above calculations are not valid when compounds of light elements such as $Be_n B_m$ are concerned.

Chemical effects on the muon capture ratio have been studied in several target systems. As this review paper does not aim at providing a complete bibliography on

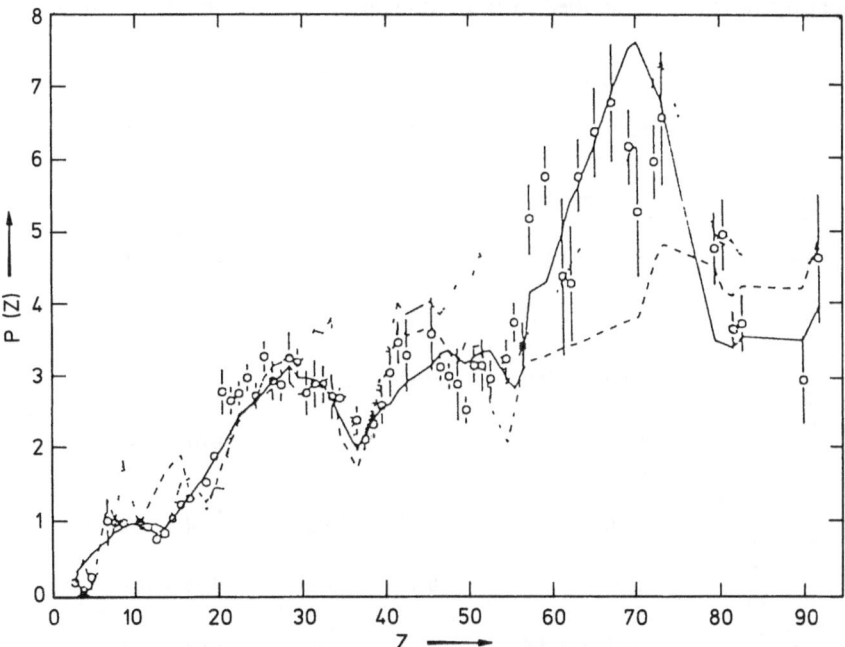

Fig. 15. Experimental and calculated capture probabilities. The solid, dashed and dotted lines correspond to the values calculated by von Egidy et al. [38], Daniel et al. [36], and Schnewly et al. [37], respectively

the capture ratio problem, only a few typical examples are cited. If the muon capture ratio in target pairs is compared with different oxidation state atoms, e.g.,

$$\overset{III}{Na}\overset{V}{NO_2}-NaNO_3, \quad Na_2\overset{IV}{S}O_3-Na_2\overset{VI}{S}O_4, \quad Na_2\overset{IV}{Se}O_3-Na_2\overset{VI}{Se}O_4,$$ muon capture occurs with higher probability for lower oxidation state atoms of the same element [40,41]. Therefore, care must be taken in evaluating the capture ratios in case of elements with different oxidation numbers.

Some uses of mesonic X-rays for analytical purposes have been published [42]. However, only in a system in which the capture ratio is well known, would a quantitative determination be practical.

3.2 Chemical Effects on Mesonic X-ray Intensity Patterns

The fact that mesonic X-ray intensity patterns are sometimes dependent on chemical environments is an interesting topic in mesonic chemistry. When the relative line intensities of mesonic X-rays are considered in such cases, systematic differences are found between different compounds of the same element. The chemical effect on muonic X-ray patterns was first discovered by Zinov et al. [43] and by Kessler et al. [44]. The results for Ti and TiO_2 [44] are shown in Table 8. It is obvious that the higher level transition to K(1s) or L(2s) displays the more prominent effect for chemical environments. Such phenomena have been observed for various systems containing sulfur [45], chlorine [45], nitrogen [41], silicon [46], vanadium [46], chromium [46], etc.

Approximately the same trend is seen for pionic X-rays. Typical examples of pionic X-ray intensity ratios are described in the following. Figure 16 shows experimental

Table 8. Intensity ratios in titanium and titanium oxide

		Ti	TiO_2	Ratio TiO_2/Ti
Lyman series	$\dfrac{3p-1s}{2p-1s}$	0.102 ± 0.003	0.093 ± 0.006	0.91 ± 0.06
	$\dfrac{4p-1s}{2p-1s}$	0.031 ± 0.004	0.027 ± 0.005	0.87 ± 0.12
	$\dfrac{5p-1s}{2p-1s}$	0.036 ± 0.003	0.025 ± 0.005	0.69 ± 0.14
	$\dfrac{6p-1s}{2p-1s}$	0.041 ± 0.003	0.028 ± 0.005	0.68 ± 0.11
	$\dfrac{higher\ K}{2p-1s}$	0.122 ± 0.007	0.062 ± 0.007	0.51 ± 0.06
Balmer series	$\dfrac{4d-2p}{3d-2p}$	0.180 ± 0.006	0.151 ± 0.006	0.84 ± 0.05
	$\dfrac{5d-2p}{3d-2p}$	0.099 ± 0.003	0.073 ± 0.006	0.74 ± 0.07
	$\dfrac{6d-2p}{3d-2p}$	0.081 ± 0.004	0.051 ± 0.006	0.63 ± 0.08
	$\dfrac{higher\ K}{3d-2p}$	0.157 ± 0.009	0.078 ± 0.016	0.50 ± 0.11

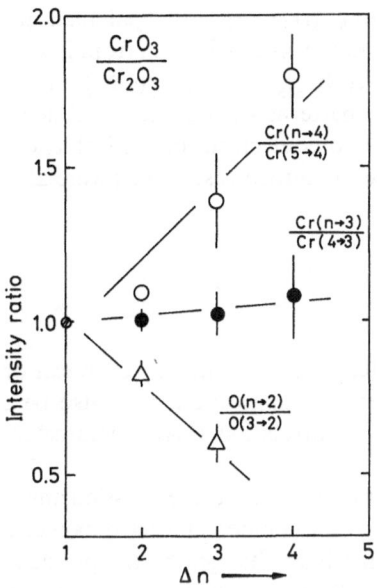

Fig. 16. Comparison of pionic X-ray intensity ratios between Cr_2O_3 and CrO_3. ○ and ● show X-rays of chromium, and Δ those of oxygen. Δn means change of the n quantum number in the same transition series

results for chromium compounds obtained by Sekine et al. [47]. Chemical influence on relative X-ray intensity ratios (CrO_3/Cr_2O_3) is observed more prominently in the transition in which the higher level with large n-quantum number is concerned. This trend is true for X-rays from both chromium and oxygen, though the latter slope is negative.

Not only the chemical constitution, but also allotropic modifications influence the X-ray intensity ratios. For carbon there are three allotropes: diamond, graphite, and soot. Experiments on muonic Lyman series intensity ratios (relative to the 2-1 transition) in these substances have been performed by Schnewly et al. [48]. The results are shown in Fig. 17. It is clear that for the higher transition, the effect of allotropic modification is more pronounced.

An explanation for the chemical effects on the mesonic X-ray intensity patterns (μ-capture by Cr in Cr_2O_3 and in CrO_3) has been given by Schnewly et al. [49] taking

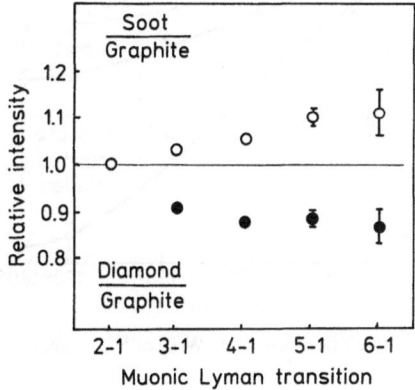

Fig. 17. Comparison of muonic X-ray intensity ratios among soot, graphite, and diamond. The transition 2-1 is taken as a standard

21

into account an initial angular momentum distribution P(l) at large principal quantum number. Naumann et al. [50] studied the K series of muonic X-rays for iron from $FeCl_3$, Fe_2O_3, $K_3Fe(CN)_6$, and aqueous solutions of $Fe(NO_3)_3$ and $K_3Fe(CN)_6$. They observed significant differences in the X-ray intensity patterns among these specimens, and discussed the importance of the structure of the outer electrons of Fe. Influences of the ionic radius and ionic charge and of hydrogen in aqueous solutions were also discussed.

4 Sum Peak Method

The sum peak method, if carefully used in the systems accompanied by chemical or biochemical change, is a useful tool for detecting such change and it may also be a convenient method to determine perturbed angular correlation (PAC) parameters without complicated electronic circuits.

When a nucleus undergoes decay and emits two γ-rays in cascade, the emission angle between the two γ-rays has a distribution depending on the mode of electromagnetic properties of the radiations. The distribution is sometimes changed by the physico-chemical or chemical environments. Figure 18 shows the decay scheme of ^{111}In which undergoes EC decay with a half-life of 2.81d and the distribution of the emission angle between two (171.3 and 245.4 eV) γ-rays in various environments. The distribution shows clear variations with the environmental conditions. This is an example of a phenomenon known as perturbed angular correlation (PAC). For nuclides such as [111m]Cd, [152]Eu, [154]Eu, and [160]Tb, PAC has been found and is applied to chemical, and sometimes, biological samples.

Sum peak formation is observed when two γ-rays emitted from a nucleus enter simultaneously into the detector. This is because cascade emission of the γ-rays in the disintegration of the nucleus is measured as one coincidence count within the detector resolution time. Figure 19 shows a γ-ray spectrum of ^{111}In which displays two single peaks of γ (171.3 and 245.4 keV) and a sum peak at 416.7 keV. The sum peak

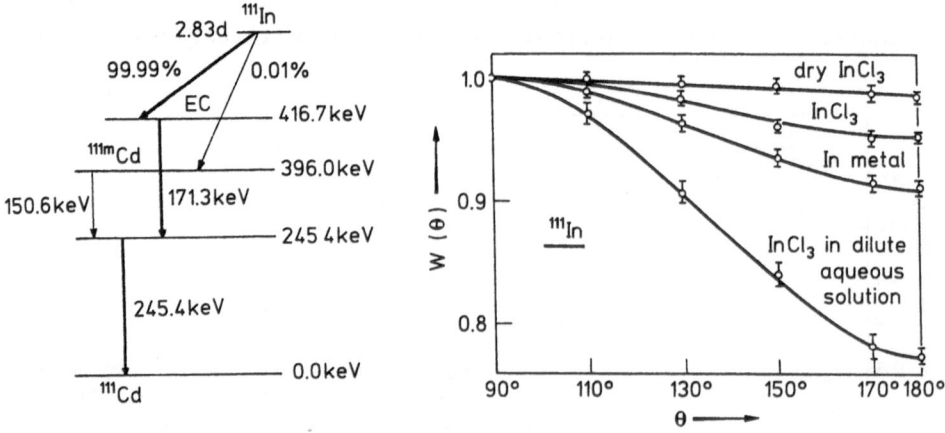

Fig. 18. Decay scheme of ^{111}In and the perturbed angular correlation functions W(θ) of ^{111}In-labeled compounds

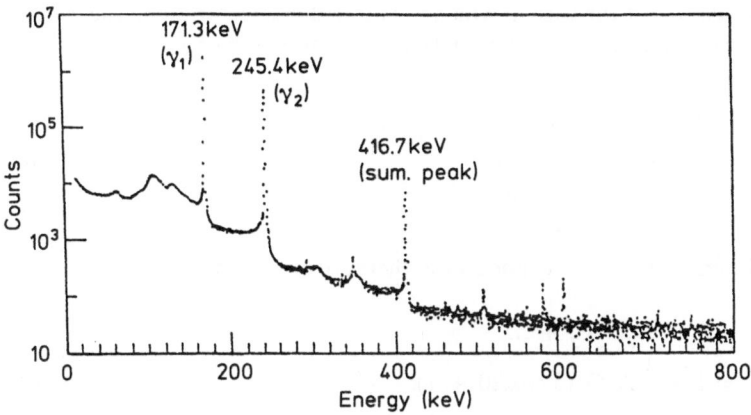

Fig. 19. γ-Ray spectrum of ^{111}In with formation of a sum peak

intensity is closely related to the phenomenon of PAC, because it orginates from the decaying nucleus which emits two γ-rays in cascade, their emission being angularly dependent in the PAC nuclides cited above. Accordingly, the intensity ratio (R) of the sum peak to the single peak varies with the environment surrounding the PAC nuclides. Conversely, the value R can be used to estimate PAC parameters if the measurement conditions are fixed for the detector system. Use of the sum peak in such ways is called the "sum peak method".

4.1 Basic Considerations on the Formation of a Sum Peak

A description of some basic considerations on the formation of a sum peak in a detector of cylindrical type may be useful. When a point source and a cylindrical detector are placed as in the geometrical arrangement in Fig. 20, the sum peak intensity ratio is given by the following relation [51]:

$$R = I_{sum}/I_{\gamma 1} = \varepsilon_{\gamma 2} \cdot \overline{W(\theta)} \cdot f \cdot k \tag{7}$$

where I_{sum} is the intensity of the sum peak; $I_{\gamma 1}$ is the intensity of the single peak of γ_1; $\varepsilon_{\gamma 2}$ is the detection efficiency for γ_2; $\overline{W(\theta)}$ is the value of the angular correlation function averaged over the angle θ; f is given by $\{W(90°) + W(180°)\}/2$; and k is a correction factor due to decrease of γ_1 by sum peak formation. This expression is based on a time integral PAC.

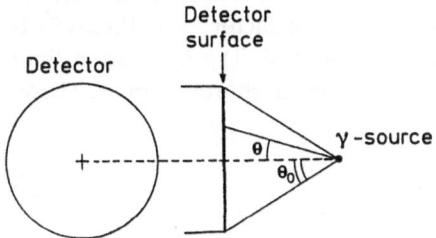

Fig. 20. Geometry of a γ-ray source placed in front of a detector

23

$\overline{W(\theta)}$ can be calculated for the case in which γ_1 enters the detector along the central axis and γ_2 enters the same detector at an angle θ to the direction of γ_1:

$$\overline{W(\theta)} = \frac{\int_0^\theta W(\theta) \sin\theta \cos^2\theta \, d\theta}{\int_0^\theta \sin\theta \cos^2\theta \, d\theta} \tag{8}$$

For ^{111}In, the following correlation function is theoretically derived:

$$W(\theta) = 1 + A_2 G_2 P_2 + \dots$$
$$= 1 + \frac{1}{4} A_2 G_2 (3 \cos 2\theta + 1) \tag{9}$$

where A_2 is the coefficient of the PAC; G_2 is the attenuation factor; and P_2 is the Legendre polynomial of second order. Therefore:

$$\left.\begin{array}{l} \overline{W(\theta)} = A/B \\ A = (160 + 64A_2 G_2) - 30(4 + A_2 G_2) \cos\theta_0 \\ \quad - 5(8 + 5A_2 G_2) \cos 3\theta_0 + 9A_2 G_2 \cos 5\theta_0 \\ B = 40(4 - 30 \cos\theta_0 - \cos 3\theta_0) \end{array}\right\} \tag{10}$$

Recently, Kudo et al. [52] proposed that when the time differential PAC function $W(\theta, t)$ is taken into account, the observed sum peak intensity I_{sum} is approximately expressed as follows:

$$I_{sum} = Ar_1 r_2 \int_\Omega \frac{\varepsilon_1'}{4\pi} \, d\omega_1 \int_\Omega \int_0^{2\tau} W(\theta, t) \, d\omega_2 \, \varepsilon_2' \, dt \tag{11}$$

where A is the activity of a nucleus; r_1, ε_1' and r_2, ε_2' are the emission ratios and the detection probabilities of the photons γ_1 and γ_2, respectively; $d\omega_1$ and $d\omega_2$ are the differential solid angles concerned with the emission derections of γ_1 and γ_2; Ω is the solid angle measured from the specimen over the detector; and $W(\theta, t)$ is the time differential PAC function:

$$W(\theta, t) = \frac{e^{-t/\tau_N}}{4\pi\tau_N} [1 + A_2 G_2(t) P_2(\cos\theta)] \tag{12}$$

where θ stands for the angle between the directions of two γ rays; t is the time interval of the emission; τ_N is the lifetime of the intermediate state; A_2 is the angular correlation coefficient; $G_2(t)$ is the time differential attenuation factor; and $P_2(\cos\theta)$ is the Legendre polynomial of second order already described. The integration limit 2τ in Eq. (11) may be replaced by infinity, because τ_N is much shorter than 2τ. Therefore:

$$I_{sum} = Ar_1 r_2 \int_\Omega \frac{\varepsilon_1'}{4\pi} \, d\omega_1 \int_\Omega W(\theta) \, \varepsilon_2' \, d\omega_2 \tag{13}$$

where $W(\theta)$ is the time integral PAC function:

$$W(\theta) = \frac{1}{4\pi}[1 + A_2\overline{G_2}P_2(\cos\theta)] \qquad (14)$$

where $\overline{G_2}$ is the time integral attenuation factor. Equation (13) is rewritten as:

$$I_{sum} = Ar_1r_2 \int_\Omega \frac{\varepsilon_1'}{4\pi} d\omega \frac{\displaystyle\int_\Omega \frac{\varepsilon_1' d\omega_1}{4\pi} \int_\Omega \frac{1}{4\pi}[1 + A_2\overline{G_2}P_2(\cos\theta)]\,\varepsilon_2'\,d\omega_2}{\displaystyle\int_\Omega \frac{\varepsilon_1'}{4\pi} d\omega}$$

$$= Ar_1r_2\varepsilon_1\varepsilon_2[1 + A_2\overline{G_2}F_{geo}] \qquad (15)$$

$$\varepsilon_1 \equiv \int_\Omega \frac{\varepsilon_1'}{4\pi} d\omega, \qquad \varepsilon_2 \equiv \int_\Omega \frac{\varepsilon_2'}{4\pi} d\omega \qquad (16)$$

$$F_{geo} \equiv \frac{\displaystyle\int_\Omega \int_\Omega \frac{\varepsilon_1'}{4\pi}\frac{\varepsilon_2'}{4\pi}P_2(\cos\theta)\,d\omega_1\,d\omega_2}{\displaystyle\int_\Omega \frac{\varepsilon_1'\,d\omega}{4\pi} \int_\Omega \frac{\varepsilon_2'\,d\omega}{4\pi}} \qquad (17)$$

The single peak counting rates I_1 and I_2 are approximately expressed as:

$$I_1 = Ar_1\varepsilon_1, \qquad I_2 = Ar_2\varepsilon_2 \qquad (18)$$

The sum peak ratios are given by:

$$\left.\begin{aligned}
R_1 &= I_{sum}/I_1 = r_2\varepsilon_2[1 + A_2\overline{G_2}F_{geo}] \\
R_2 &= I_{sum}/I_2 = r_1\varepsilon_1[1 + A_2\overline{G_2}F_{geo}]
\end{aligned}\right\} \qquad (19)$$

Conversely, the value of $\overline{G_2}$ can be estimated by:

$$\left.\begin{aligned}
\overline{G_2} &= \frac{1}{A_2F_{geo}}\left(\frac{R_1}{r_2\varepsilon_2} - 1\right) \\
&= \frac{1}{A_2F_{geo}}\left(\frac{R_2}{r_1\varepsilon_1} - 1\right)
\end{aligned}\right\} \qquad (20)$$

25

Since the observed sum peak intensity I_{sum}^{obs} always contains the contribution from accidental coincidence of two different nuclei emitting cascade γ rays, the evaluation of the sum peak ratios should be carried out according to the following relations:

$$\left.\begin{aligned}\frac{I_{sum}}{I_1} &= \frac{I_{sum}^{obs}}{I_1} - 2\tau I_2 \\[2mm] \frac{I_{sum}}{I_2} &= \frac{I_{sum}^{obs}}{I_2} - 2\tau I_1\end{aligned}\right\} \tag{21}$$

These basic relations described above can be utilized in application of the sum peak method to chemical and/or biological samples.

4.2 The Sum Peak Method as Applied to Chemistry

De Bruin et al. [53] found that in an aqueous solution the sum peak intensity ratio of ^{181}Hf is slightly influenced by F^- concentration. Since ^{181}Hf is one of the PAC nuclides, their finding was explained by a change in the angular distribution of two γ's due to F^- complex formation. The magnitude of the effect is only of the order of 2%. They used an NaI(Tl) detector with resolution much lower than that of a Ge(Li) or pure germanium detector. Probably, in such a measurement system it is not easy to detect a 2% difference with high reliability. No subsequent reports have been published using an NaI(Tl) detector.

When a Ge(Li) detector of high resolution is used under proper conditions, dependence of the sum peak intensity ratio on chemical environments can be made much more remarkable. The ratio $I_{sum}/I_{\gamma 245}$ for ^{111}In is definitely dependent on the source-detector distance. It is true that counting statistics are still significant at a distance of 100 mm at which geometrical conditions are good enough to detect a sum peak intensity change. If the source is brought much closer to the detector, the change will be too obscured. On the other hand, if the source is too far from the detector, counting statistics become worse.

Table 9. Sum peak intensity ratios for ^{111}In

| | Material | Sum peak intensity ratio | | |
		observed	calculated	corrected[a]
$\dfrac{\gamma_{sum}}{\gamma_{247}}$	InCl$_3$ (solid)	0.0429 ± 0.0007	0.0334	0.0351
	In metal	0.0389 ± 0.0005	0.0319	0.0335
	InCl$_3$ (aq)	0.0366 ± 0.0008	0.0294	0.0309
	In · edta (solid)	0.0427 ± 0.0011		
	In · edta (aq)	0.0361 ± 0.0016		
$\dfrac{\gamma_{sum}}{\gamma_{171}}$	InCl$_3$ (solid)	0.0309 ± 0.0006	0.0257	0.0270
	In metal	0.0295 ± 0.0003	0.0247	0.0259
	InCl$_3$ (aq)	0.0265 ± 0.0006	0.0227	0.0238
	In · edta (solid)	0.0312 ± 0.0008		
	In · edta (aq)	0.0262 ± 0.0012		

[a] corrected for accidental sum peak formation

The sum peak intensity ratios measured for ^{111}In sources in different states of indium chloride (in the solid and in an aqueous solution) and in metallic form are shown in Table 9. It is clear that the ratios are dependent on the source states. For the same sample under the same geometrical conditions, the sum peak intensity ratios was calculated using Eq. (7). The trend of calculated values is in agreement with that for measured values as listed in Table 9. Difference in the absolute values between measured and calculated ratios may be partly due to nonuniformity of efficiency inside the detector and partly due to accidental coincidence, etc.

Yoshihara [54] measured ^{160}Tb in terbium oxide Tb_4O_7 and in terbium chloride solution by the 'sum peak pair method', and pointed out that the difference of the sum peak intensity ratio between two states could be enlarged by using comparison of a sum peak pair. This method has an advantage that the dependence of the sum peak pair ratio on the source-detector distance is not as noticeable as that in the ordinary sum peak method.

4.3 The Sum Peak Method Applied to Biological Substances

The PAC method can be applied to biological substances; information obtained by this method is sometimes so unique and valuable that other methods cannot compete with it. The sum peak method based on the principle of PAC is also expected to provide useful information on biological samples.

The first experiment in this direction was performed by Yoshihara et al. [55] with radioactive europium (^{152}Eu + ^{154}Eu) added to a solution of bovine serum albumin (BSA). The sum peak intensity ratio I_{sum}/I_{1408} of the original acid solution of ^{152}Eu is clearly changed by incorporation in the BSA solution. The ratio recovered its original value when BSA was destroyed by 6 N hydrochloric acid. This experiment confirms that the effect of metal ions on biological macromolecules can be examined by the sum peak method.

The above experimental result is further verified by Yoshihara et al. [56] using bovine serum albumin and ^{111}In. A scheme of the chemical procedure is shown in Fig. 21. An aqueous solution of indium chloride at pH 3.6 is added to an aqueous solution of bovine serum albumin (pH 5.6). Then, by addition of 0.1 N hydrochloric acid the metal-BSA combination is destroyed. The samples 1, 2, and 3 are measured by a γ-ray spectrometer. The sum peak intensity ratios obtained by the measurement are

Table 10. Sum peak intensity ratios for each step of chemical treatment in the ^{111}In-BSA system

Case	Sample No.[a]	I_{sum}/I_{171}	I_{sum}/I_{245}
A	1	0.0411 ± 0.0013	0.0557 ± 0.0011
with	2	0.0435 ± 0.0009	0.0627 ± 0.0022
detector 1	3	0.0414 ± 0.0006	0.0585 ± 0.0011
B	1	0.0495 ± 0.0005	0.0697 ± 0.0035
with	2	0.0551 ± 0.0081	0.0754 ± 0.0060
detector 2	3	0.0499 ± 0.0006	0.0704 ± 0.0033

[a] See chemical procedure in Fig. 2

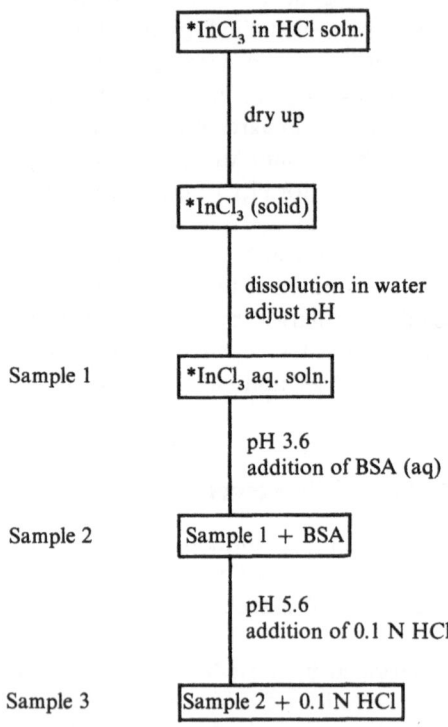

Fig. 21. Chemical procedure for labeling bovine serum albumin (BSA) with ^{111}In and destruction of In-BSA

listed in Table 10. The value for sample 2 is different from those of samples 1 and 3. This shows the formation of the metal-BSA combination in sample 2. The combination of ^{111}In-BSA is further verified chemically by gel chromatographic behavior of samples 1, 2, and 3 as shown in Fig. 22. The chromatographic peak of ^{111}In-free ion in sample 1 is shifted to the higher molecular weight direction (smaller elution volume), showing labeling of BSA with ^{111}In in sample 2. The peak position in sample 3 is again the same as that of ^{111}In-free ion in sample 1.

Application of this sum peak method to the study of human platelet cells has appeared in recent papers of Kudo et al. [52, 57]. The procedure for labeling the human platelets is just as follows: About 43 ml of blood were drawn into a syringe containing 7 ml of ACD-A solution (sodium citrate 2.2 w/v%, pH 4.5 ~ 5.5). Then, the blood was centrifuged and the supernatant platelet-rich plasma (PRP) was transferred to a tube in which it was ajusted to pH 6.5 by adding ACD-A solution. The platelets were sedimented by centrifuging and resuspended in ACD-A saline solution containing 11.1 MBq of ^{111}In-tropolone at pH 6.5. The mixture was incubated for 20 min at room temperature. About 15 ml of platelet-poor plasma (PPP) was added to the mixture, followed by centrifugation to encourage platelet formation. The ^{111}In-labeled platelets were resuspended in 10 ml PPP to remove contaminating red cells. Finally, centrifugation gave clean ^{111}In-labeled platelets.

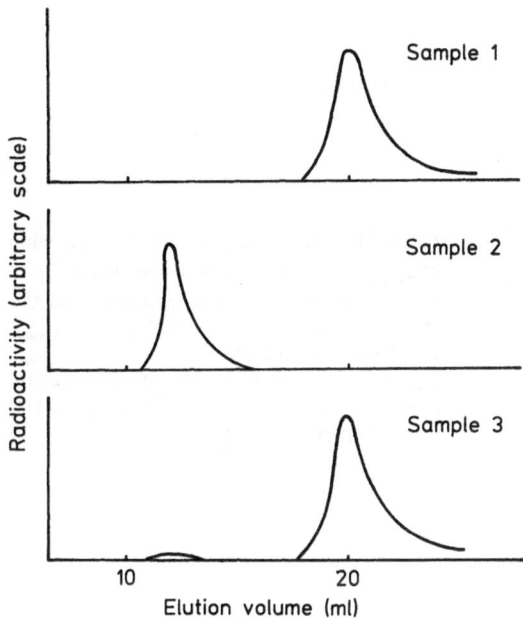

Fig. 22. Elution curves of the samples 1, 2, and 3. The sample numbers correspond to those in the procedure shown in Fig. 21

Typical results for the ^{111}In-labeled platelets are shown in Table 11 together with control ^{111}In^{3+} free ions. The platelets show higher sum peak intensity ratios compared to those of the control. Moreover, the value is dependent on the groups (a) and (b). Probably this means that the state in (a) is different from that in (b) for either biochemical or medicochemical considerations.

Kudo et al. [52] studied the sum peak intensity ratios of ^{111}In-labeled samples involving bovine serum albumin, human platelet cells, indium chloride, and indium tropolone complex. From the measured value of the sum peak intensity ratio, the PAC parameter $\overline{G_2}$ was calculated using Eq. (20). The results are shown in Fig. 23. Both the ratios for 171 keV γ and 245 keV γ give straight lines when they are plotted against $\overline{G_2}$ for various samples. It is interesting to note that the different specimens of human platelet cells (b), (b)′, and (b)″ show different $\overline{G_2}$. This may be due to the difference in environments of ^{111}In which reflect the difference of material components in the cells of individuals.

Table 11. Relative sum peak intensity ratios for ^{111}In-labeled platelet

Specimen	Relative sum peak intensity ratio		
	$\dfrac{I_{sum}}{I_{171}} \Big/ \left(\dfrac{I_{sum}}{I_{171}}\right)_{control}$	$\dfrac{I_{sum}}{I_{245}} \Big/ \left(\dfrac{I_{sum}}{I_{245}}\right)_{control}$	average
In-platelet (a)	1.100 ± 0.027	1.108 ± 0.064	1.104 ± 0.038
control (In^{3+} aq)	1.000	1.000	1.000
In-platelet (b)	1.216 ± 0.031	1.230 ± 0.046	1.223 ± 0.027
control (In^{3+} aq)	1.000	1.000	1.000

[a] Different specimens which may reflect different states

Kenji Yoshihara

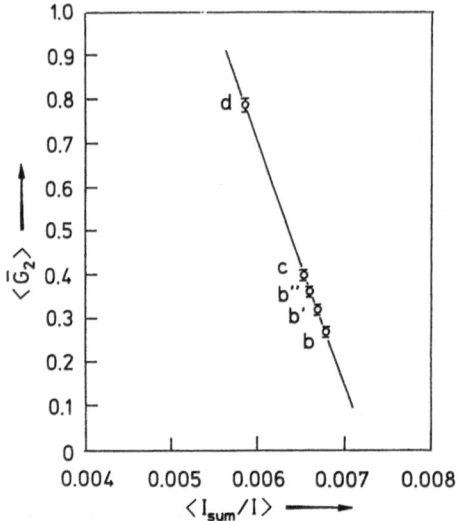

Fig. 23. $\overline{G_2}$ calculated from the observed value of I_{sum}/I using Eq. (20). The averaged $\langle \overline{G_2} \rangle$ is the average of the $\overline{G_2}$'s determined from the intensity ratios I_{sum}/I_{171} and I_{sum}/I_{245}. $\langle I_{sum}/I \rangle$ means the average of I_{sum}/I_{171} and I_{sum}/I_{245} for samples: b, b', b'' (human platelet cells), c (saline solution of $^{111}InCl_3$), and d (^{111}In-tropolone complex)

These authors recommend the following procedure to determine the PAC parameter $\overline{G_2}$:

1) Determine the efficiencies ε_1 and ε_2 and a geomettical factor F_{geo} using Eqs. (20) and (21) by use of some samples whose $\overline{G_2}$ are known.
2) Measure the γ-ray spectra of the specimen in question and calculate the sum peak intensity ratios according to Eq. (21).
3) Compute the $\overline{G_2}$ value according to Eq. (20).

Although the sum peak method does not allow the $\overline{G_2}$ values to be determined as accurately as in ordinary PAC measurements, it has some advantages over the PAC method. Only an aliquot of labeled substances is required for the measurement, the measurement is much shorter, and the apparatus needed is much simpler. Future development of this method should be encouraged.

5 Conclusion

Photon intensity ratio variations due to chemical environments are interesting phenomena for detailed study.

These phenomena have possible applications in material analysis. Work in this direction has begun. Although there are still many problems to be solved, application of these phenomena is promising, as seen in the examples cited in the preceding sections.

However, the phenomena occurring in the systems are usually complicated, and still require basic research. Complete theories to explain them are not available at present. Nevertheless, the many features of these phenomena will certainly increase interests of researchers in basic and applied fields.

6 References

1. Scofield JH (1974) Phys. Rev. A 9: 1941
2. Arndt E, Brunner G, Hartmann E (1982) J. Phys. B: At. Mol. Phys. 15: L887
3. Yoshihara K, Lazzarini E (1984) X- and γ-ray spectrum peak intensity change due to chemical environments, in: Hot Atom Chemistry, Matsuura T (ed) p 348, Amsterdam Tokyo, Elsevier Kodansha
4. Tamaki Y, Omori T, Shiokawa T (1975) Radiochem. Radioanal. Lett. 20: 255
5. Tamaki Y, Omori T, Shiokawa T (1978) Jpn. J. Appl. Phys., Suppl. 17: 425
6. Lazzarini E, Lazzarini-Fantola AL, Mandelli-Bettoni M (1978) Radiochim. Acta 25: 81
7. Tamaki Y, Omori T, Shiokawa T (1979) Radiochem. Radioanal. Lett. 37: 39
8. Yoshihara K, Hibino A, Yamoto I, Kaji H (1981) Radiochem. Radioanal. Lett. 48: 303
9. Yamoto I, Kaji H, Yoshihara K (1985) J. Radioanal. Nucl. Chem., Lett. 95: 301
10. Yamoto I, Kaji H, Yoshihara K (1986) J. Chem. Phys. 84: 522
11. Mazaki H (1978) J. Phys. E 11: 739
12. Johannsen B, Münze R, Dostal KP, Nagel M (1981) Radiochem. Radioanal. Lett. 47: 57
13. Bainbridge KT, Goldhaber M, Wilson E (1951) Phys. Rev. 84: 1260
14. Gerasimov VN, Zelenkov AG, Kulakov VM, Pchelin VA, Soldatov AA, Stepanchikov VA, Chistyakov LV (1981) Soviet J. Nucl. Phys. 34: 1
15. Gerasimov VN, Zelenkov AG, Kulakov VM, Pchelin VA, Sokolovskaya MV, Soldatov AA, Chistyakov LV (1982) Soviet Phys. JETP 55: 205
16. Kiss K, Pálinkás J, Schlenk B (1980) Radiochem. Radioanal. Lett. 45: 213
17. Izawa G, Fujiwara H, Omori T, Yoshihara K, Arai H, Sera K, Ishii K (1987) J. Radioanal. Nucl. Chem. Lett. 118: 59
18. Yoshihara K, Fujiwara H, Iihara J, Izawa G, Omori T, Arai H, Sera K, Ishii K (1988) Application of Ion Beams in Materials Science, Sebe T (ed) p. 531, Tokyo, Hosei Univ. Press
19. Benka O (1980) J. Phys. B: At. Mol. Phys. 13: 1425
20. Kawatsura K, Ozawa K, Fujimoto F, Terasawa M (1977) Phys. Lett. 60A: 327
21. Ozawa K, Kawatsura K, Fujimoto F, Terasawa M (1976) Nucl. Instr. Methods 132: 517
22. Hopkins F, Little A, Cue N, Dutkiewicz V (1976) Phys. Rev. Lett. 37: 1100
23. Benka O, Watson RL, Bandong B (1983) Phys. Rev. A 28: 3334
23a. Uda M, Benka O, Fuwa K, Maeda K, Sasaki Y (1987) Nucl. Instr. Methods B22: 5
24. Brunner G, Nagel M, Hartmann E, Arndt E (1982) J. Phys. B: At. Mol. Phys. 15: 4517
25. Mukoyama T, Kaji H, Yoshihara K (1986) Phys. Lett. A 118: 44
26. Band IM, Kovtun AP, Listengarten MA, Trzhaskovskaya MB (1985) J. Electron Spectrosc. Relat. Phenom. 36: 59
27. Mukoyama T, Taniguchi K, Akachi H (1986) Phys. Rev. B 34: 3710
28. Johnson KH (1973) Adv. Quantum Chem. 7: 143
29. Adachi H, Tsukada M, Satoko C (1978) J. Phys. Soc. Jpn. 45: 875
30. Hartmann E, Der R, Nagel M (1979) Z. Phys. A 290: 349
31. Hartmann E, Arndt E, Brunner G (1980) J. Phys. B: At. Mol. Phys. 13: 2109; (1987) ibid. 20 175
32. Collins KE, Heitz C, Cailleret J (1980) J. Chem. Research (S) 263; (1980) (M) 3401
33. Collins KE, Collins CH, Heitz C (1981) Radiochim. Acta 28: 7
34. Tamaki Y, Omori T, Shiokawa T (1982) Bull. Miyagi Unv. Educ. 17: 1
35. Fermi E, Teller E (1947) Phys. Rev. 72: 349
36. Daniel H (1979) Z. Phys. A 291: 29
37. Schnewly H, Pokrovsky VI, Ponomarev LI (1978) Nucl. Phys. A 312: 419
38. von Egidy T, Jakubassa-Amundsen DH, Hartmann FJ (1984) Phys. Rev. A 29: 455
39. Imanishi N, Furuya T, Fujiwara I, Shinohara A, Kaji H, Iwata S (1985) Phys. Rev. A 32: 2584
40. Schnewly H, Dubler T, Kaeser K, Robert-Tissot B, Schaller LA, Schellenberg L (1978) Phys. Lett. 66A: 188
41. Dubler T, Kaeser K, Robert-Tissot B, Schnewly H (1976) Phys. Lett. 57A, 325
42. Köhler E, Bermann R, Daniel H, Ehrhart P, Hartmann FJ (1981) Nucl. Instr. Methods 187: 563
43. Zinov VG, Konin AD, Mukhin AI, Polyakova RV (1967) Sov. J. Nucl. Phys. 5: 420
44. Kessler D, Anderson HL, Dixit MS, Evans HJ, McKee RJ, Hargrove CK, Barton RD, Hincks EP, McAndrew JD (1967) Phys. Rev. Lett. 18: 1179

31

45. Mausner LF, Naumann RA, Monald JA, Kaplan SN (1977) Phys. Rev. A *15*: 479
46. von Egidy T, Denk W, Bergmann R, Daniel H, Hartmann FJ, Reidy JJ, Wilhelm W (1981) Phys. Rev. A *23*: 427
47. Sekine T, Hashimoto K, Kaji H, Yoshihara K (1989) J. Radioanal. Nucl. Chem., Lett. *135*: 207
48. Schnewly H, Boschung M., Kaeser K, Piller G, Rüetschi A, Schaller LA, Schellenberg L (1983) Phys. Rev. A *27*: 950
49. Hild M, Kaeser K, Piller G, Schnewly H (1985) J. Phys. B: At. Mol. Phys. *18*: 2093
50. Naumann RA, Daniel H, Ehrhart P, Hartmann FJ, von Egidy T (1985) Phys. Rev. A *31*: 727
51. Yoshihara K, Kaji H, Mitsugashira T, Suzuki S (1983) Radiochem. Radioanal. Lett. *58*: 9
52. Kudo T, Tsuchihashi N, Yui T, Mitsugashira T, Kaji H, Yoshihara K (1988) Appl. Radiat. Isot. *39*: 131
53. De Bruin M, Korthoven PJM (1975) Radiochem. Radioanal. Lett. *21*: 287
54. Yoshihara K (1983) Radiochem. Radioanal. Lett. *58*: 25
55. Yoshihara K, Kaji H, Shiokawa T (1979) Inorg. Chim. Acta *32*: 143
56. Yoshihara K, Omori T, Kaji H, Suzuki Y, Mitsugashira T (1983) Radiochem. Radioanal. Lett. *58*: 17
57. Kudo T, Tsuchihashi N, Yui T, Mitsugashira T, Kaji H, Yoshihara K (1987) Appl. Radiat. Isot. *38*: 123

7 Note Added in Proof

Chemical effects of X-ray intensity ratios have been studied in a series of niobium compounds by electron and proton bombardments [58]. Molecular orbital calculation using the GAMESS method has been applied to estimate MO occupation of the bonding electrons. The $L_{\gamma 1}/L_{\beta 1}$ X-ray intensity ratio linearly increased with increase of the MO occupation of niobium. The relation between the intensity ratio and the occupation is considered to be essentially important for describing the chemical effects, although a similar relation holds good between the ratio and ionicity as a gross trend. (Fig. A)

The sum peak method was studied to determined a PAC parameter $\overline{G_{22}}$ of ^{111}In in hydrochloric acid [59]. It is dependent on pH, especially in a 4–14 region. At higher concentrations of hydrochloric acid, however, $\overline{G_{22}}$ stays at a constant value inspite of formation of an $InCl_6^{3-}$ type complex.

A semi-theoretical approach was introduced to estimate a PAC parameter $\overline{G_{22}}$ using specimens with finite dimension instead of a point source [60].

The sum peak method was applied to estimate K electron capture probabilities (P_k) by Singh and Sahota. They determined P_k values of 437 and 384 keV levels of ^{133}Ba [61,62]. The method consists of measuring a sum peak of K X-rays emitted following K capture (or K conversion) and successively emitted γ-rays in the energy spectrum. The observed values were in good agreement with those obtained by other authors and with theoretical ones. The advantages of this method are: firstly it is independent of the K-shell fluorescence yield ω_k and secondly there is no need to calculate the absolute K X-ray detection efficiency. They also determined P_k values for Eu 172, 103 and 97 keV levels after EC decay of ^{153}Gd [63] by this method. They extended the sum peak method to estimate the ω_k values of ^{75}As produced from ^{75}Se decay [64]. It is obvious that their method is available when chemical effects are not expected or negligible. The method was applied to K capture probabilities in the decay of ^{175}Hf [65].

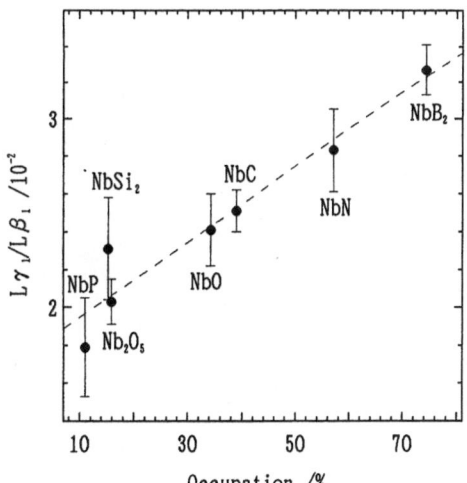

Fig. A. Relation between the $L_{\gamma 1}/L_{\beta 1}$ X-ray intensity ratio and MO occupation in electron bombarded niobium compounds

References

58. Iihara J, Izawa G, Omori T, Yoshihara K (1990) Nucl. Instr. Methods Phys. Research B (in press)
59. Kaji H, Yoshihara K (1987) J. Radioanal. Nucl. Chem., Lett. *119*: 143
60. Kudo T, Tsuchihashi N, Mitsugashira T, Yoshihara K (1987) J. Radioanal. Nucl. Chem., Lett. *119*: 131
61. Singh K, Sahota HS (1983) J. Phys. G. Nucl. Phys. *9*: 1565
62. Singh K, Sahota HS (1983) J. Phys. Soc. Jpn. *52*: 2336
63. Singh K, Grewal BS, Sahota HS (1985) J. Phys. G: Nucl. Phys. *11*: 399
64. Singh K, Sahota HS (1984) J. Phys. G: Nucl. Phys. *10*: 241
65. Singh K, Gill TS, Singh K (1988) J. Phys. Soc. Jpn. *57*: 3762

Radiometric Determination of Trace Elements

Nobuo Suzuki

Department of Chemistry, Faculty of Science, Tohoku University, Sendai 980, Japan

Table of Contents

Many radioanalytical methods have been introduced for trace analysis. Each method has its own advantages, but in comparison with ordinary instrumental nuclear methods, these radioanalytical methods, except for substoichiometry, have not been applied to analysis of many actual samples. Substoichiometric analytical methods including substoichiometric radioactivation analysis will come to be more widely used, as a result of the many possible applications.

Apart from the careful regulation of radioactive material handling, the main obstacle to more widespread application of these radioanalytical methods is that they involve chemical procedures for sample preparation. By examining the chemical procedures carefully, simpler methods may be found as has been demonstrated in substoichiometric radioactivation analysis and some cases of substoichiometric chemical speciation.

Topics in Current Chemistry, Vol. 157
© Springer-Verlag Berlin Heidelberg 1990

1 Introduction

Many nuclear techniques have been introduced into the field of analytical chemistry. This chapter focuses on radioanalytical methods with the exception of radiotracers, used to correct for separation yield in analytical procedures, and the radioreagent technique, based on quantitative and stoichiometric reaction of an element of interest. Instrumental methods such as radioactivation analysis and proton-induced X-ray emission spectrometry are also powerful nuclear analytical techniques, but they also must fall outside the scope of this short treatment. Then, the modern trends in radio-analytical methods are described, which feature autonomous analytical methods with a novel combination of chemical procedures and radioisotope techniques.

2 Radioanalytical Methods

2.1 Substoichiometric Analysis

The principle of the substoichiometric analysis is as follows [1]: To the element of interest (M_x), one adds a known amount of its labeled radioisotope with the specific activity $S = A/M$, where A is the radioactivity and M the amount of carrier; hence the specific activity of the mixture becomes $S' = A/(M_x + M)$. Knowing the change in specific activity from S to S', the element of interest can be simply determined. This is the same as the principle of isotope dilution, but in practice the accurate determination of S and S' is very tedious. This is one important reason why isotope dilution analysis is not very popular in trace analysis. However, in substoichiometric analysis, equal amounts of the element (m) are isolated substoichiometrically from the radioisotope solution and the mixed solution, and subsequently the radioactivities of the separated portions (a and a') are measured, then the amount of the element of interest can be calculated according to the equation:

$$M_x = M\{(S/S') - 1\} = M\left\{\left(\frac{a}{m}\right)\middle/\left(\frac{a'}{m}\right) - 1\right\}$$

hence: $M_x = M\{(a/a') - 1\}$ (1)

This means that the element of interest can be determined simply by measuring the radioactivity. The substoichiometric determination procedure consists of three steps: radioisotopic labelling of the element of interest, reproducible separation of a fraction of the element of interest, and measurement of the radioactivity of the separated portion. In the first step, it is necessary to achieve an isotopic equilibrium between the element of interest and the added radioisotope. The second step, known as substoichiometric separation, is the most important one. The absolute amount of the separated portions is not of concern; a constant amount of the element of interest must be separated with high reproducibility. The substoichiometric separation is usually made by an appropriate chemical separation method. In solvent extraction, for example, this is made by adding a substoichiometric amount of reagent which is less than the amount that would be required for complete recovery of all the element of interest. In the third step, conventional techniques used for radioactivity measurements can be applied. Using Eq. 1, the element of interest can be determined from just the radioactivity measurement of the substoichiometrically separated portion which is one significant advantage of substoichiometry [2].

The present concept can easily be combined with radioactivation analysis. The specific activity of the element of interest in an irradiated unknown sample changes after adding a known amount of carrier M, hence leading to similar forms of Eq. 1. This is based on reverse isotope dilution, but by applying substoichiometry, an unknown amount of element can be determined only by the radioactivity measurement without any comparison with a standard sample. Substoichiometry can be best applied to usual radioactivation analysis based on a comparison of the radioactivities from sample and standard. A test sample containing an unknown amount of the element of interest M_x and a standard sample containing M_s are simultaneously irradiated. In this case, larger but equal amounts of carrier (M) are added to the test and the standard samples after irradiation. The element of interest can be determined by simply comparing the radioactivities (a_x and a_s) of the substoichiometrically separated portions from the test and the standard samples. The relevent equation is shown as:

$$A_x/(M_x + M) \approx A_x/M = a_x/m$$

and

$$A_s/(M_s + M) \approx A_s/M = a_s/m$$

hence:

$$M_x = M_s(a_x/a_s) \tag{2}$$

In this method, substoichiometric separation is applied in the presence of a large amount of carrier, hence the substoichiometric isolation is quite easy and selectivity between interferring diverse elements is much improved.

2.2 Sub- and Super-equivalence Method

This is a method based on the measurement of interphase distribution of a substance such as the reaction product between the element of interest and an appropriate reagent. In this case the amount of reagent used is not always a substoichiometric amount. The substance to be determined is labeled with an appropriate isotope, and two series of aliquots are taken [3]. In the first series, each aliquot contains the same amount of the substance to be determined, x. In the second series, each aliquot contains an amount k times greater than the first, $x' = kx$. The first series of aliquots is isotopically diluted by adding incremental amounts of the nonradioactive form of the substance to be determined y. This gives a series of incremental isotopically diluted aliquots, $x + y_1, x + y_2, \dots x + y_n$. To all aliquots is added the same amount of the reagent, and the radioactivity of the product is measured after an appropriate interphase distribution of the product. The amount of product generally will depend on the concentration of the element of interest in the aliquot. In the second series, the same amount of element m_0 with the same radioactivity a_0 is isolated from all the aliquots. From the first series, unequal amounts of the element $m_1, m_2, ... m_n$ with radioactivity $a_1, a_2, ... a_n$ are isolated. This leads to:

$$\frac{a_0}{a} = \frac{m_0}{m}\left(1 + \frac{y}{x}\right) \tag{3}$$

The concentration of the substance in one of the aliquots of the first series will be equal to that of the second series, where $(x + y_i)/v_0 = kx/v_0$. Curve 1 in Fig. 1 means isolation of exactly the same amount of substance from all aliquots and curve 2 means isolation of the total amount of substance to be determined from the aliquots. The working region described lies between curves 1 and 2. Generally, all curves have a common intersection at their inflection points. The intersection point is independent of the amount of reagent and can be used for the determination of the substance of interest as demonstrated in Fig. 1.

The errors associated with this method when using a solvent extraction system have been theoretically discussed [4]. The error decreases distinctly with increasing k, and k values in the range 2–4 can be recommended. The sub- and super-equivalence method allows determination at a concentration level two or three orders of magnitude lower than substoichiometric isotope dilution analysis. Originally the sub- and super-equivalence method was introduced for determining radioactively labeled substances, but of course this method can be applied to determine a nonactive substance as a reverse isotope dilution method [5]. Solvent extraction is a popular technique to measure the interphase distribution of the substance but other general techniques such as precipitation reactions have also been used [6]. Analytical examples using the sub- and super-equivalence method are limited, with most being model demonstrations to examine the characteristics of the method.

From the theory of the sub- and super-equivalence method of isotope dilution analysis, it follows that all real curves of isotope dilution $a_0/a_i = f(y)$ must cross an

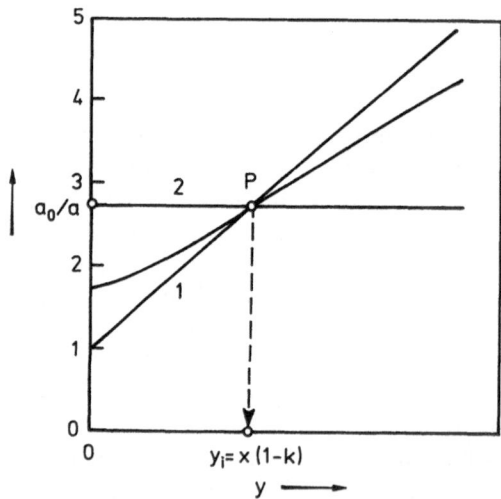

Fig. 1. Dependence of the isolated activity ratio on the added known amount

isoconcentration point P. However, near the isoconcentration point P, this function can be considered as quasi-linear. After the preliminary determination of the element of interest by a usual graphic interpretation, a new restricted analysis is carried out by the least square method. This duplicate analysis allows a simple determination of the isoconcentration point P and lower the determination error. This has been demonstrated in the determination of 0.112 µg Zn(II)/ml in a model radioactive solution [7], and it has been applied to the determination of zinc in engine oil [8].

2.3 Concentration-Dependent Distribution Method

This radiochemical method is based on the utilization of calibration curves showing the dependence of the distribution ratio of the element of interest in a two-phase system on the total concentration of the element [9]. The concentration-dependent distribution method is applied to the simultaneous determination of chemically similar elements. This method uses a competitive reaction and is based on calibration curves obtained by two substoichiometric systems for a known and fixed total concentration of the two similar elements. In system (1), the concentration of element A serves as the independent variable and the concentration of element B is a constant parameter for each curve. In system (2) the opposite is used. The substoichiometric separation is carried out in both systems and calibration curves are drawn. The shape of the calibration curve has been theoretically derived and the influence of experimental conditions on the determination error and optimum labeling substances, yielding minimal error, have been discussed in connection with extraction constants [10, 11].

It is possibile to expand this idea to the simultaneous determination of three or four chemically similar elements, but the use of calibration curves and graphical treatment cannot be applied to more than two elements without much loss of simplicity and clarity. The sub- and super-equivalence method is superior to the concentration-dependent distribution method in that it requires less sample preparation for radioactivity measurements and fewer measurements. On the other hand, the calibration curves of the sub- and super-equivalence method are generally steeper than those of the concentration-dependent distribution method, which would favor the latter.

3 Substoichiometric Separation

Substoichiometric separation is performed by ordinary chemical separation methods such as solvent extraction, ion exchange, precipitation, and electrochemical methods. In recent years, however, the ion exchange and electrochemical methods have not been used very much in substoichiometric separation. The precipitation technique is often used due to its simplicity, while solvent extraction is most widely employed. This is because the procedure for solvent extraction is very simple and an appropriate extraction system can usually be selected from the great number of research papers dealing with solvent extraction of many different elements. Two extraction systems are commonly used: chelate extraction of metal ions with chelating agents; and ion-association extraction of metal ions with simple negative or positive ions.

3.1 Precipitation

Substoichiometric precipitation of sulfur as barium sulfate has been applied to the determination of sulfur in coal [12]. The experimental procedure involves the combustion of a coal sample labeled with ^{35}S in a flowing oxygen stream, collection of the combustion gases in hydrogen peroxide, and substoichiometric precipitation of barium sulfate. The precipitate is collected on a membrane filter, and then its radioactivity is measured by a GM counter. Samples of 0.2–10.0% sulfur are accurately determined with $\pm 2\%$ RSD by measuring the activity of the precipitates thus obtained and with comparison to standard samples subjected to the same procedures. Analysis times of 15 minutes per sample are typical.

The substoichiometric precipitation of fluorine has been examined with lanthanum fluoride. Fluoride ion labeled with ^{18}F is precipitated substoichiometrically with lanthanum as shown in Fig. 2. Fluorosilicate is also precipitated substoichiometrically. This procedure was applied for the determination of oxygen in silicon crystals [13]. The nuclear reaction used is $^6Li(n, \alpha)t$, $^{16}O(t, n)^{18}F$. Upon irradiation of the test and standard samples in a nuclear reactor, the samples are dissolved. After addition of NaF, some of the fluoride is precipitated with a substoichiometric amount of lanthanum. By comparing the activity of the substoichiometric precipitates for standard and test samples, oxygen content can be calculated by Eq. 2. The substoichiometric precipitation of barium with sulfate was studied and applied to the determination of traces of uranium, where ^{140}Ba, produced by nuclear fission of ^{235}U, was used [14]. After irradiation of test and standard samples, barium carrier is added. Barium is preliminary separated as barium carbonate, then dissolved in HCl. Precipitation of barium with a substoichiometric amount of sulfate is reproducible over a wide range of pH values. Uranium content can be determined from the radioactivities of the substoichiometric precipitates from the test and standard samples.

A reproducible coprecipitation or adsorption reaction of a constant amount of the element of interest can be used in substoichiometry. An interesting example is the substoichiometric radioactivation analysis for oxygen, based on the reproducible isolation of fluorine or fluorosilicate with a substoichiometric amount of hydrated tin dioxide [13]. It has been applied to the determination of oxygen in silicon crystal.

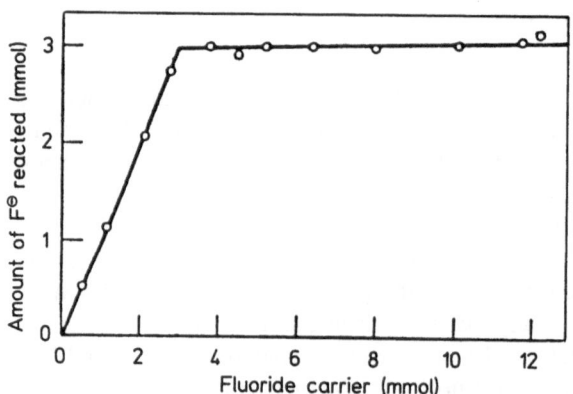

Fig. 2. Reproducibility of the substoichiometric precipitation of fluoride with lanthanum

The test and standard samples are irradiated in a nuclear reactor. After dissolution in the presence of 12 mmol fluorine carrier, the solution is adjusted to pH 8–9. The hydrated silicon dioxide precipitate is filtered and the acidity of the filtrate is adjusted to $1M$ nitric acid. Then it is passed through a column packed with hydrated tin dioxide. The radioactivity of the column is measured. The oxygen content is calculated from the activities of the test and standard samples. This method is easy and quick, and is suitable for determinations of trace amounts of oxygen.

Another example shows the possibility of using coprecipitation of the element of interest in the presence of a substoichiometric amount of complexing agent. In solutions containing between 10^{-3}–$10^{-5}M$ zirconium, a substoichiometric amount of ethylenediaminetetraacetic acid (EDTA) is added to both the sample and the standard solutions, labeled with ^{95}Zr. The uncomplexed zirconium is completely isolated from the solution by coprecipitation with copper(II) or iron(II) hydroxide. Zirconium determination in the concentration range of 10^{-4}–10^{-6} g/ml in simulated mixtures and steel samples has been demonstrated [15].

3.2 Solvent Extraction

Substoichiometric extraction of an ion associate is sometimes preferable to that of internal complex compounds, although the latter form very stable compounds in many cases. The substoichiometric isolation of Mn(VII), Au(III), Ta(V), Re(VII), and other elements in the form of ion associates with tetraphenylarsonium (TPA) or quarternary ammonium base (QAB) is undoubtedly the most expedient from the viewpoint of selectivity.

Conditions of the substoichiometric extraction of the ion associates as shown in Eq. 4 has been theoretically discussed [16].

$$(m - n) KA_0 + (MA_m)^{m-n} \rightleftharpoons K_{m-n}MA_{m,0} + (m - n) A^- \qquad (4)$$

KA is the original salt of a bulky cation with a complex anion of metal and the subscript o denotes the organic phase. When a substoichiometric amount of the extractant has been added, which is 50% of the theoretical equivalent amount to total MA_m^{m-n}, 99% portion of this extractant is used up in forming the ion associate. These conditions lead to:

$$C_M = \frac{200}{K_{ex}^{1/m-n}(m - n)} \left(\frac{1}{\beta[A]^n} + [A]^{m-n} \right)^{1/m-n} \qquad (5)$$

Calculated conditions for substoichiometric extraction of chloride anionic complexes of cobalt, copper, zinc, and cadmium with Aliquot-336 (trioctylmethylammonium chloride) or other quaternary ammonium salts have been compared with experimental results.

The substoichiometric extraction of fluoride from 50% dimethylsulfoxide (DMSO) aqueous phase via a substoichiometric replacement reaction with triphenyltin chloride into chloroform was examined. Reproducible substoichiometric separations of fluoride in the presence of bromide, chloride, iodide, nitrate, thiocyanate, or sulfate were demonstrated [17].

The substoichiometric extraction of phosphorus was studied, in which phosphorus was extracted in the form of a ternary compound such as ammonium phosphomolybdate, 8-hydroxyquinolinium phosphomolybdate, tetraphenylarsonium phosphomolybdate, and tri-n-octylamine phosphomolybdate [18]. In the application of these phosphomolybdate compounds to the substoichiometric determination of phosphorus, two types of substoichiometric extraction were examined: one in which phosphorus separation was achieved by adding a substoichiometric amount of molybdenum, and the other in which a substoichiometric amount of organic reagent was added. The former is applicable to ion associates of phosphomolybdate with four organic reagents. The extraction of ammonium phosphomolybdate is best for substoichiometric determination of phosphorus. Phosphorus in orchard leaves (NBS SRM-1571) and spinach (SRM-1570) was determined by various substoichiometric methods after an addition of ^{32}P or radioactivation [19]. In these comparisons ammonium phosphomolybdate with a definite amount of molybdenum was extracted into MIBK. The analytical data agree well with the respective certified value given by NBS.

The appropriate extraction condition of the metal ion with a substoichiometric amount of a chelating agent HA was discussed from the basic equation of the overall extraction process as:

$$M^{n+} + nHA_0 \rightleftharpoons MA_{n0} + nH^+ \tag{6}$$

The corresponding equilibrium constant given below is defined as the extraction constant:

$$K = \frac{[MA_n]_0 [H^+]^n}{[M^{n+}] [HA]_0^n} \tag{7}$$

As contrasted with an ordinary extraction using excess chelating agent, in the substoichiometric extraction, almost all the chelating agent, say 99.9%, is used up in forming the chelate MA_n. Hence, by substituting the equilibrium concentrations of $[MA_n]_0$, $[HA]_0$, and $[M]$ at the substoichiometric condition, the threshold pH is given by:

$$pH \geq \frac{1}{n} \log \left(\frac{C_{HA}}{n} \right) - \frac{1}{n} \log \left(C_M - \frac{C_{HA} \cdot V_0}{n \cdot V} \right)$$

$$- \frac{1}{n} \log K - \log (0.001 C_{HA}) \tag{8}$$

where V is the volume of the phases and C the original concentration [20]. It follows from an analysis of Eq. 8 that the first two terms on the right-hand side have a comparatively small effect on the value of the threshold pH. The last two terms have a much greater effect. In general, the threshold pH value of substoichiometric extraction shifts to higher pH values than in ordinary extraction using an excess amount of reagent. Careful selection of the extraction system with a higher extraction constant is very important to guarantee quantitative extraction in the lower threshold pH region.

Selectivity of the substoichiometric extraction must be compared with that of an ordinary extraction. The extraction selectivity of two metal ions $M(n^+)$ and $M'(m^+)$ with a chelating agent is shown as the ratio of the metals in the extracted organic phase:

$$\frac{[MA_n]_0}{[M'A_m]_0} = \frac{K[HA]^{n-m}[M]}{K'[H]^{n-m}[M']}.$$ (9)

In the simplest case where charge and concentration of M are equal to those of M', the quantitative separation of M from M' is considered as:

$$[MA_n]_0/[M'A_n]_0 > 100 \quad \text{and} \quad [M]/[M'] < 0.01$$

Hence, in the extraction with an excess amount of reagent the ratio K/K' should be greater than 10^4 to fulfill the above requirement. For substoichiometric extraction using the substoichiometric amount of reagent to extract one half of the metal of interest, quantitative separation $[MA_n]_0/[M'A_n]_0 > 100$ can be achieved in the extraction system for which the ratio K/K' is greater than 200 because M/M' equals 0.5. Higher selectivity of the substoichiometric extraction has been experimentally demonstrated in the substoichiometric extraction of mercury with dithizone, and theoretically a large excess of metals with lower extraction constants (silver, copper, bismuth, cadmium, cobalt, nickel, iron, lead, zinc, manganese, and thallium) have been shown not to interfere, and this was one of the advantages of the present extraction system when applied to mercury determination in actual samples [21, 22]. Dithizone, diethyl-dithiocarbamate, 8-hydroxyquinoline, N-benzoyl-N-phenylhydroxylamine and many other chelating agents have been used for substoichiometric chelate extraction. Examples have been summarized in a number of review articles [23, 24, 25].

Substoichiometric separation by solvent extraction of the extractable portion of the element of interest after masking it by a substoichiometric amount of complexing agent such as EDTA and related complexing agents presents another interesting possibility for substoichiometric techniques. Substoichiometry using two chelating agents, a complexon and an extracting agent, offers a method in which a water-soluble stable complex is formed with a substoichiometric amount of complexon, while the unmasked element of interest is separated by solvent extraction. This method is based on the differences in reactivity of complexon and extracting agent with the element of interest. Since complexons react with many elements including rare earth and rare elements, and form high stability complexes, their use in the substoichiometric separation extends the applicability of solvent extraction and increases the selectivity offered by substoichiometric determination techniques.

The optimum conditions for the substoichiometric separation of the element by using two chelating agents have been calculated from the following equation [26−29]:

$$\alpha_A[H] \left\{ \frac{10^{-3} K_{MY} C_{H_jY}}{a_j K_{ex}} \right\}^{1/n} \geq C_{HA} \geq \alpha_A[H] \left(\frac{10^3}{K_{ex}} \right)^{1/n}$$ (10)

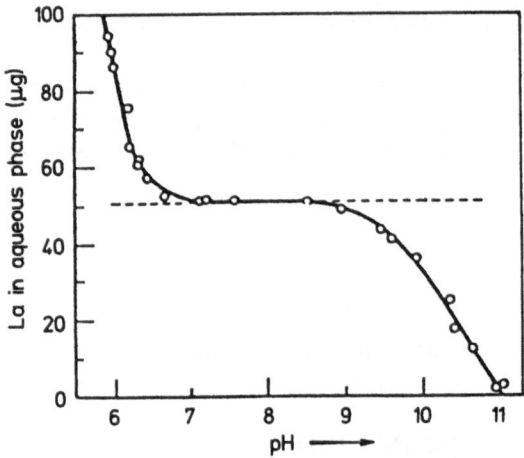

Fig. 3. Effect of pH on the substoichiometric separation of lanthanum

where H_jY denotes the complexon; MY, the water soluble complex; HA, the extracting agent; K_{MY}, the thermodynamic stability constant; K_{ex}, the extraction constant
of M with HA; C, the initial concentration; n, the reaction ratio between M and HA;
a_j, a factor which determines the quantitative distribution of any species of the complexon; and α_A, a factor expressed as $1 + 1/P_{HA} + K_a/(P_{HA}[H])$ where P_{HA} and K_a
are the partition coefficient and the dissociation constant, respectively. The effect
of pH on the substoichiometric separation of lanthanum is shown in Fig. 3 [29]. In
this case, half the lanthanum is complexed with EDTA. Except for the low and high
pH regions, where a preferential complexation with EDTA or with oxine occurs, the
flat plateau indicates that the substoichiometric separation of lanthanum can be
performed in this region. This plateau region from pH 7.3 to 8.5 is in good agreement
with the optimum pH region calculated theoretically. This method has been applied
to the determination of lanthanum by radioactivation analysis. The samples and
lanthanum standard were irradiated in a nuclear reactor. Lanthanum carrier was added
to the irradiated materials. After preliminary separation by precipitation and/or
solvent extraction, the pH of the lanthanum solution thus obtained was adjusted to
a value greater than pH 9.5. Lanthanum was extracted with $0.25M$ 8-hydroxyquinoline
in chloroform. The organic phase was shaken with $3.0 \times 10^{-5}M$ EDTA solution and
lanthanum was separated substoichiometrically into the aqueous phase, and its
radioactivity was compared with that obtained from the standard. Analytical results
for 1.4 ppm La in orchard leaves (NBS-SRM-1571) and 320 ppb La in spinach (SRM-
1570) are in good agreement with the reported values. The present method offers
simple and reliable detection of trace amounts of lanthanum.

Substoichiometry with two chelating agents, a complexon and an extracting agent,
can be applied to the simultaneous determination of a series of metals [30]. The principle
is that the element which has the lowest stability constant with a complexon is separated
substoichiometrically by an appropriate extraction, while other elements which have
higher stability constants with the complexon are completely masked and separated
quantitatively. This substoichiometric group separation was combined with the
neutron activation analysis for some rare earth elements. The optimum conditions
for the substoichiometric separation of lanthanum, europium, and terbium with

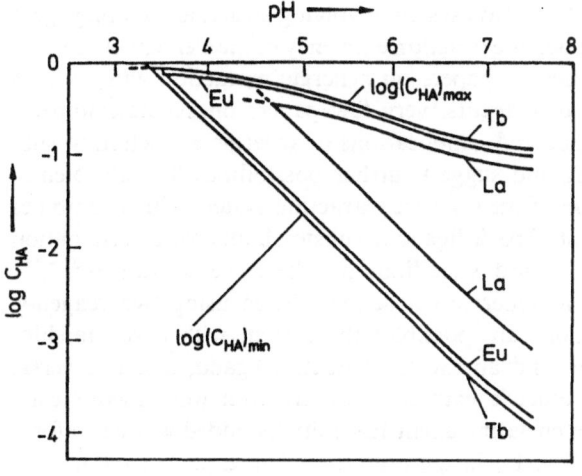

Fig. 4. Optimum conditions for the substoichiometric separation of lanthanum, europium, and terbium using DTPA and TTA; $x = 0.5$, $C_{DTPA} = 1 \cdot 10^{-4} M$

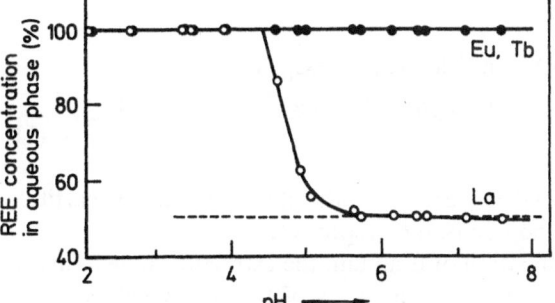

Fig. 5. Effect of pH on the simultaneous separation of lanthanum, europium, and terbium by the substoichiometric separation of lanthanum. La $= 6.28 \cdot 10^{-5} M$, Eu $= 3.70 \cdot 10^{-7} M$, Tb $= 3.78 \cdot 10^{-7} M$, TTA $= 0.1 M$

diethylenetriaminetetraacetic acid (DTPA) and thenoyltrifluoroacetone (TTA) can be calculated by Eq. 10 as shown in Fig. 4. The regions enclosed by the two curves $\log C_{HA, min}$ and $\log C_{HA, max}$ contain the optimum conditions at X = 0.5 (50% substoichiometry). The wider region for europium and terbium compared with the case of lanthanum shows the preferential complexation of these elements with DTPA. Figure 5 shows the effect of pH on a simultaneous separation in the presence of a substoichiometric amount of DTPA. On the lanthanum curve, the flat plateau, where 50% of the lanthanum remains in the aqueous phase, indicates that the substoichiometric separation of lanthanum can be performed, and in this region, europium and terbium are quantitatively separated to the aqueous phase. This method has been applied to the simultaneous determination of lanthanum, europium, and terbium in orchard leaves. After neutron irradiation, lanthanum, europium, and terbium carriers were added to both the orchard leaves and the standard. The amounts of europium and terbium carriers were 200 times smaller than that of lanthanum. Preliminary separation by fluoride precipitation, hydroxide precipitation, and di(2-ethylhexyl)phosphoric acid (HDEHP) extraction was carried out to isolate these rare earth elements. The final aqueous solution was subjected to substoichiometric separation of lanthanum with DTPA and TTA. Analytical results of 1.24 ppm La, 25 ppb Eu, and 17 ppb Tb are obtained.

Nobuo Suzuki

It is well known that a synergic extraction system involving an acidic chelating agent and a neutral basic ligand provide better extraction efficiency of the element of interest. In contrast to the number of research papers on synergic extraction with various combinations of different types of reagents, very few papers on substoichiometry using synergic extraction have appeared. Applications of synergic extraction to sub-stoichiometry are very interesting and suggest further possibilities for substoichiometric separation. The application of the synergic extraction system which combines an acidic chelating agent and a neutral basic ligand to substoichiometric determination of calcium, uranium, manganese, and vanadium has been demonstrated [31-35]. In substoichiometric extraction of synergic extraction system using two reagents, two substoichiometric combinations are possible: the system with a substoichiometric amount of chelating agent and an excess of neutral ligand, and vice versa. Both systems for the substoichiometric extraction of uranium were theoretically discussed [32]. When $100x\%$ of the chelating agent HA initially added is used to form the extractable adduct complex MA_nL_m in the presence of an excess of the neutral ligand L, the pH of the aqueous phase can be readily calculated from the synergic extraction constant $K_{ex,s}$:

$$pH = n^{-1} \log (xC_{HA}/n) - n^{-1} \log (C_M - xC_{HA}V_0/nV)$$
$$- \log [(1 - x) P_{HA}C_{HA}/(1 + P_{HA})] - n^{-1} \log C_A - n^{-1} \log K_{ex,s}$$

$$(11)$$

where C, V, P, and $K_{ex,s}$ are the initial concentration, the phase volume, the partition coefficient, and the synergic extraction constant, respectively.

For the substoichiometric determination of uranium, the extraction system involving a substoichiometric amount of TTA and an excess of tributylphosphate (TBP) was compared with a system involving a substoichiometric amount of trioctylphosphine oxide (TOPO) and an excess of TTA. As an example, an extraction curve of U(VI) in the former system is shown in Fig. 6. This extraction curve is highly consistent with the theoretical extraction curve computed by Eq. 11. A constant amount of U(VI) is extracted at pH 5.1–7.0, and this plateau pH region is wider than that of the extraction system involving a substoichiometric amount of TOPO and an excess of

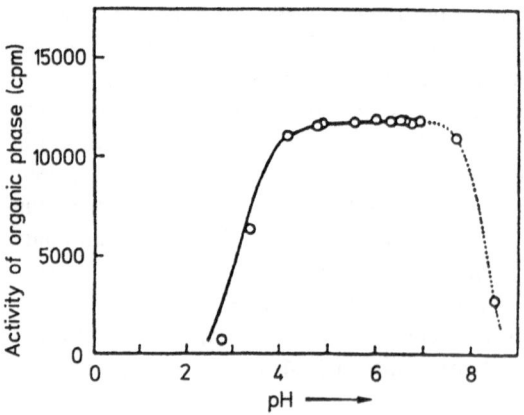

Fig. 6. Effect of pH on the substoichiometric extraction of $2.6 \times 10^{-5} M\ UO_2^{2+}$ with $2.0 \times 10^{-5} M$ TTA and $0.18 M$ TBP. Solid line is the theoretical curve (see text)

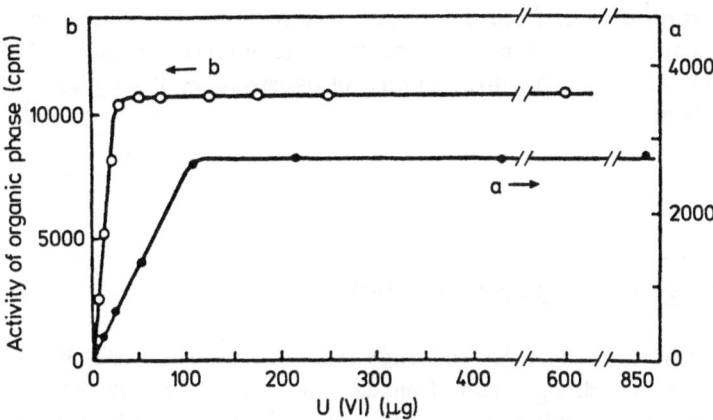

Fig. 7. Reproducibility of the substoichiometric extraction.
(a) $4.6 \times 10^{-5}M$ TOPO, $2.3 \times 10^{-3}M$ TTA; pH 2.5–3.6,
(b) $2.0 \times 10^{-5}M$ TTA, $0.18M$ TBP; pH 5.8–6.4

Table 1. Precision of the substoichiometric extraction

Substoichiometry with respect to neutral ligand[a]		Substoichiometry with respect to chelating agent[b]	
U(VI) taken (μg/10 ml)	Activity of organic phase (cpm)	U(VI) taken (μg/10 ml)	Activity of organic phase (cpm)
238	4814	48	10707
286	4845	72	10702
476	4787	120	10694
571	4843	168	10723
714	4854	240	10725
952	4871	599	10828
	Mean = 4836 ± 30 cpm		Mean = 10730 ± 50 cpm
	R.s.d. = 0.62%		R.s.d. = 0.46%

[a] $4.5 \times 10^{-5}M$ TOPO, $2.3 \times 10^{-3}M$ TTA, pH 2.7–3.1, shaking time, 10 min.
[b] $2.0 \times 10^{-5}M$ TTA, $0.18M$ TBP, pH 6.1–6.4, shaking time, 2 h.

TTA. Substoichiometric extraction was applied to a series of solutions containing various quantities of U(VI) labeled with ^{237}U. As shown in Fig. 7, the constant radio-activity of the organic phases guarantees a high reproducibility of these substoichiometric systems. Table 1 lists the reproducibility of the two substoichiometric extraction systems. A high reproducibility under 1% relative standard deviation (RSD) is possible.

The accuracy and precision of the substoichiometry were evaluated by determining U(VI) in a simulated mixture containing a known amount of U(VI) and large amounts of diverse ions together with phosphoric acid. After a preliminary extraction of U(VI) from $6M$ nitric acid into 2% TBP in toluene, the substoichiometric extraction was applied. The high precision of 0.23% RSD and the high accuracy of +0.5% relative

47

deviation of the determined value from the added amount of U(VI) are very satis-factory [32]. Another combination of hexafluoroacetylacetone (HHFA) and TOPO can be applied to more complicated matrix samples of phosphate rock samples containing lower amounts of U [35].

4 Modifications

4.1 Isotopic Exchange and Replacement Methods

An analytical method based on isotopic exchange was introduced in the 1950s. A few papers on the isotopic exchange method include the exchange system between metal and metal species, for example silver and solid silver chloride [36] and silver and silver dithizonate in a liquid-liquid extraction system [37]. In a direct isotopic exchange, the inactive element with unknown weight W_x can be determined by adding a definite amount of its labeled species with a known weight W_1 and radioactivity A_1:

$$\frac{A_1}{W_x + W_1} = \frac{A_2}{W_1} \tag{12}$$

where A_2 is the radioactivity of labeled species át the exchange equilibrium; therefore:

$$W_x = W_1 \left(\frac{A_1}{A_2} - 1\right) \tag{13}$$

This method, however, is restricted in that the isotope exchange reaction of concern should proceed rapidly to attain the exchange equilibrium. Therefore, the methods cited above cannot be applied for rather slow isotope exchange systems before the exchange equilibrium is attained. A modified isotopic exchange method based on analysis of the exchange rate of slow, but measurable, exchange reactions has been introduced [38]. For an isotopic system with a measurable exchange rate

$$AX + BX^* \rightleftharpoons AX^* + BX$$

the following McKay's equation holds:

$$\ln(1 - F) = \frac{W_1 + W_x}{W_1 W_x} Rt \tag{14}$$

where:

$$F = At/A_\infty = \frac{At}{W_x \cdot A(W_1 + W_x)}$$

and AX denotes the compound of X to be analyzed; BX, the compound of X to be added; F, fraction of exchange at time t; W_x, weight of X in AX; W_1, weight of X in

BX; R, rate of the exchange reaction; A initial activity of BX; At, activity of AX at time t; and A_∞, activity of AX at the exchange equilibrium. By measuring the activities A_1 and A_2 of the AX fraction at times t_1 and t_2, X can be obtained by the following equation:

$$X = \frac{m}{\{(2A_1 - A_2)/A_1^2\}\,(A-1)}$$

(15)

Thus, by choosing t_2 to be twice as large as t_1, X can be readily obtained even before the exchange equilibrium is attained.

The determination of iodine in organic iodine compounds by this non-equilibrium isotopic exchange method has been demonstrated, where an acetone or ethyl alcohol solution of organic iodine compound was mixed with potassium iodide solution spiked with ^{131}I. After an appropriate mixing time, an aliquot of the reacting mixture was tranferred into an ice-cooled two phase solution of benzene and water. On shaking vigorously, any further isotopic exchange reaction is quenched. Isotopic exchange curves are shown in Fig. 8. The results for various amounts of n-butyl iodide are summarized in Table 2. Concentrations in the range of 6.5–635 mg I can be determined with an error of $\pm 4\%$.

Replacement of the metal of interest with labeled metal sulfide has also been report-ed [39-41]. Metal determinations by replacement of this metal with partially labeled metal sulfide has been introduced, in which metal in the surface layer of metal sulfide was labeled with radioisotope [42]. For the preparation of partially labeled zinc sulfide, as an example, zinc sulfide was prepared by an ordinary method and thoroughly washed free of sulfide. Then tracer ^{65}Zn solution was added and equilibrated for 3 h. By adding different amounts of copper in the range 5–25 µg to a constant amount of

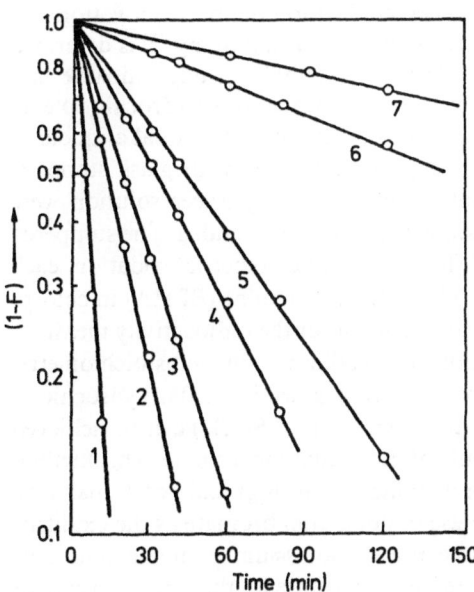

Fig. 8. Isotope-exchange curves. Curve 1: CH_3I, 2: C_2H_5I, 3: n-C_3H_7I, 4: n-C_4H_9I, 5: n-$C_6H_{13}I$, 6: iso-C_4H_9I, 7: CH_2I_2

Table 2. Determination of iodine in n-C_4H_9I

n-C_4H_9I taken, mg I	KI* added, mg I	Temperature, °C	n-C_4H_9I found, mg I	Error, %
635 (0.5M)	635	35	654	+3.0
100	100	35	102	+2.0
		45	104	+4.0
10.0	100	35	9.77	−2.3
		45	9.98	−0.2
6.90	100	30	6.71	−2.8
		45	7.09	+2.8
	10.0	45	6.67	−3.3
1.00	100	45	1.09	+9.0
0.69	10.0	50	0.73	+5.8

labeled zinc sulfide, quantitative displacement of copper with ^{65}Zn was verified by measuring the released activity. A replacement method based on the solvent extraction technique offers wider ranges of concentrations for the labeled species, specific activities, and experimental conditions, such as pH. A displacement method for the determination of micro-amounts of Hg(II) has been developed, where ^{65}Zn in zinc-1(2-pyridylazo-2-naphthol) complex was quantitavely displaced with Hg(II) at pH 5. In this method, 10–80 µg of mercury can be determined [43].

4.2 Redox Substoichiometry

Substoichiometric oxidation or reduction followed by a chemical separation such as coprecipitation or solvent extraction has been introduced [44, 45]. Redox substoichio-metry combined with solvent extraction has been applied to both radioactivation and isotope dilution analyses [46, 47]. Redox substoichiometry for antimony [48] is described here. First, all the Sb(V) was reduced to Sb(III) by bubbling sulfur dioxide gas. Arsenic(III) was removed by extraction with benzene into a 10M hydrochloric acid solution. After adjusting the acidity of the aqueous acidic solution, equal aliquots of the solution were placed into brown-colored glass test tubes with glass stoppers. Increasing amounts of Sb(III) carrier solution and ^{125}Sb(III) tracer solution were added and then 0.1 ml of $4.0 \times 10^{-4} N$ potassium dichromate was added. The stoppered tubes were allowed to stand for 120 min. After the substoichiometric oxidation, each solution was shaken with 0.05M N-benzoylphenylhydroxylamine (BPHA) in chloro-form and an aliquot of the aqueous phase was taken out for the radioactivity measure-ment. To apply the antimony determination to metallic tin, the substoichiometric oxidation of Sb(III) in the presence of Sn(IV) was investigated in a 1.0M hydrochloric acid solution. As shown in Fig. 9, the constant oxidation of Sb(III) can be achieved by the addition of substoichiometric amounts of potassium dichromate. This method was applied to antimony determination in metallic tin (10 mg) and good analytical results of 1.22 ± 0.05 µg are obtained. The use of potassium bromate as the oxidizing agent was discussed in comparison with the use of potassium permanganate and potassium dichromate [49]. Potassium dichromate has advantages that the quantitative

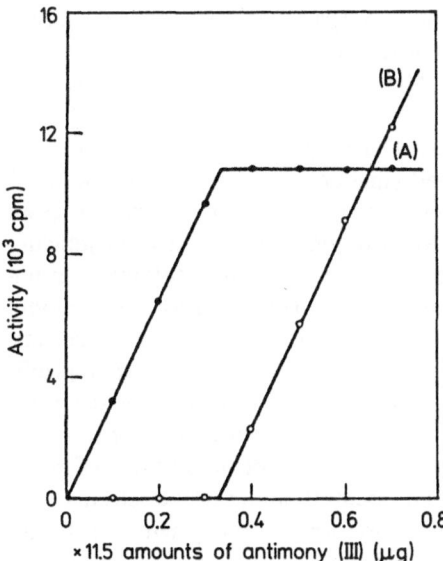

Fig. 9. Substoichiometric oxidation of antimony(III) by adding 1.0 ml of $6.0 \cdot 10^{-5} N$ potassium dichromate in the presence of 10 mg of tin(IV): (A) aqueous phase; (B) organic phase

oxidation of Sb(III) is not disturbed by the presence of zinc, its dilute solution is very stable compared with that of permanganate, and the oxidation of Sb(III) amounts, as small as 2 µg and even in the presence of zinc, proceeds to completion at room temperature. This means that the redox substoichiometry can be applied to the determination of very small amounts of antimony.

Redox substoichiometry can be combined with radioactivation analysis, and this was applied to the antimony determination in metal samples [46, 50]. After dissolution and acidity adjustment of the irradiated metal sample, all Sb(V) was reduced to Sb(III), and quite a large amount of Sb(III) carrier was added. A substoichiometric amount of Sb(III) was oxidized by the addition of potassium bromate, and the solution was shaken with BPHA to extract unoxidized Sb(III). An aliquot of the organic phase was taken out for the radioactivity measurements. A standard containing a known amount of antimony was treated in the same way. Antimony in metallic zinc samples (reagent grade) was determined by redox substoichiometric isotope dilution analysis and redox substoichiometric activation analysis with consistent results [50].

4.3 Substoichiometric Chemical Speciation

Almost all publications on substoichiometry which have appeared to date have dealt with the determination of the total amount of the element. Recently, some attention has been focused on the determination of the element of interest in different chemical states. This speciation of trace elements of interest is almost impossible by instrumental analytical methods alone. For example, suppose a sample contains the element of interest in different oxidation states, abbreviated as M(III) and M(V). If one state, e.g., M(III) can be determined, the total amount of M can be ascertained after appropriate reduction of M(V) to M(III) in the sample. Then the amount of M(V) can be estimated as the difference in the total amount of M and that of M(III) in sample.

This approach is generally used in conventional analysis, but when it is considered that actual samples sometimes contain several different chemical states of the element of interest, the independent determination of M(III), M(V), and other species is most desirable.

Selective substoichiometry has been proposed and applied to the determination of antimony in different oxidation states [51], methylmercury [52], and butyl tin species [53, 54]. The principle is based on the substoichiometric separation of the element of interest in a particular chemical state from a sample. An example of the selective substoichiometry for As(III) and As(V) is as follows: to the sample solution containing an unknown amount M_x of As(III) and another unknown amount of As(V), a known amount of radioactive As(III) with a specific activity $S = A/M$ is added. Then the specific activity of As(III) becomes $S' = A/(M_x + M)$. A substoichiometric separation specific for As(III) is applied and the radioactivity of the substoichiometrically separated portion is measured. The amount of M(III) can be determined according to Eq. 1. The As(V) can be determined by the same procedure, except for the addition of radioactive As(V) instead of radioactive As(III). The essential points of the selective substoichiometry are selection of the substoichiometric separation system specific for As(III) and for As(V), and ensuring that no isotopic exchange or electron transfer between As(III) and As(V) can proceed.

For the substoichiometric separation of As(III) and As(V), a chelate extraction system using thionalide and an ion-pair extraction system using pyrogallol together with tetraphenylarsonium are appropriate, respectively. The effect of acidity on the extraction of As(III) and As(V) was thoroughly examined by varying the standing time after the addition of a thionalide solution in ethanol [55]. After standing for 30 min, the extraction of As(III) into 1,2-dichloroethane was quantitative at $3-5M$ sulfuric acid and pH 6–8, but no extraction of As(V) with thionalide was observed under these conditions. Reproducibility of the substoichiometric extraction of As(III) was examined. To centrifugal tubes containing various amounts of ^{74}As(III) solutions in $5M\ H_2SO_4$ a constant but substoichiometric amount of thionalide solution was added; then, the mixtures were allowed to stand for 30 min before extraction into 1,2-dichloroethane. The relative standard deviation of the substoichiometric extraction from five test solutions with the same component is as high as 0.9%. The selectivity of the present extraction system was examined by the substoichiometric extraction from test solutions containing both As(III) and As(V), and a constant radioactivity was obtained in the substoichiometric extractions from mixed solutions of ^{74}As(III) and As(V), but no activity was observed from mixed solutions of As(III) and ^{74}As(V). These results show clearly that no isotope exchange occurs between As(III) and As(V), and that the complete substoichiometric separation of As(III) from As(V) is achieved.

The ion-pair extraction system with pyrogallol and tetraphenylarsonium is an appropriate system for the selective sybstoichiometry of As(V) [56]. Tetraphenylarsonium was chosen as the substoichiometric agent in the presence of a large excess of pyrogallol. To examine the reproducibility of substoichiometric extraction of As(V), the substoichiometric extraction with $99.6 \times 10^{-6}M$ TPA in the presence of $0.2M$ pyrogallol was applied to a series of solutions (5 ml) containing various amounts of radioactive As(V). Figure 10 shows that the radioactivity of the organic extract increases with increasing amounts of As(V); but beyond the equivalence point of

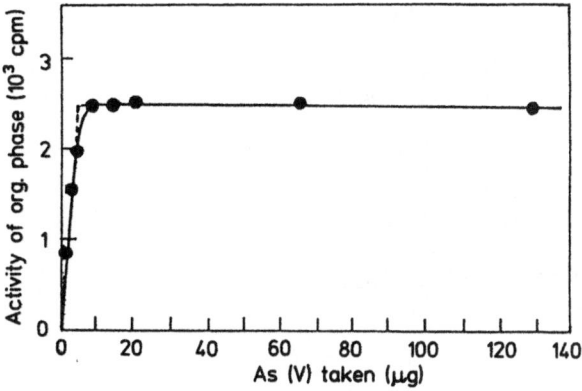

Fig. 10. Reproducibility of substoichiometric extraction of As(V). [TPA], $9.96 \times 10^{-6}M$; [pyrogallol], $0.2M$; [H$_2$SO$_4$], $1.6M$; Shaking time, 5 min

Table 3. Selectivity of substoichiometric extraction of As(V)[a]

Diverse arsenic species added/mg		Activity of substoichiometric extract/cpm
None		6173
As(III)	3.87	6017
MMA	9.97	6158
DMA	5.53	6353

Organic phase: $9.96 \times 10^{-6}M$ TPA, aqueous phase: $0.2M$ pyrogallol in $1.6M$ sulfuric acid.
[a] As(V) taken, 29.10 µg/5 ml.

TPA, a substoichiometric amount of As(V) is constantly extracted. The reproducibility of these substoichiometric extractions is very high; the relative standard deviation (RSD) of the organic extracts in the plateau region of Fig. 10 is 0.52%, and As(V) at about the 0.8 µg/ml level can be accurately determined under the cited substoichiometric conditions. The selectivity of the present method for As(V) was investigated by substoichiometric extraction from aqueous solution containing 29.1 µg of radioactive As(V) in the presence of a large amount of non-active arsenic species such as As(III), monomethylarsonic acid (MMA), and dimethylarsinic acid (DMA). The radioactivities of the substoichiometric extracts obtained in the presence or absence of diverse arsenic species are shown in Table 3. No significant interference is observed in the presence of large amounts of As(III), MMA, and DMA. The present substoichiometric extraction system is highly selective for As(V). In addition, these results demonstrate that no isotopic exchange takes place between radioactive As(V) and non-active As(III), MMA, or DMA. The presence of many different common ions did not cause any interference.

Nobuo Suzuki

Table 4. Arsenic(V) in acid digested solution of seaweed sample*

No.	Conditions of additional conc. H_2SO_4 treatment	Determined value of As(V)/μg g^{-1}
1	—**	5.89 ± 0.34
2	170 °C, 2 h	10.5 ± 0.05
3	170 °C, 20 h	65.1 ± 1.1
4	240 °C, 2 h	69.9 ± 1.9

* *Laminaria religiosa* Miyabe, Arsenic content 67.1 ± 1.7 μg/g^{-1}.
** Oridinary acid digestion with HNO_3.

The present method was applied to the determination of As(V) in an acid-digested solution of a seaweed sample. Seaweed is composed of soft tissues and easily decomposed by an ordinary acid digestion with concentrated nitric acid. The arsenic(V) content in the acid-digested solution was determined by substoichiometry as only 5.89 μg g^{-1} (Table 4), as compared with the total arsenic content of 67.1 ± 1.7 μg g^{-1} determined by a nondestructive photon activation analysis. The effect of additional acid digestion with concentrated sulfuric acid after nitric acid digestion was examined. As Table 4 shows, the determined value of As(V) increases with increased heating temperature and heating time, and these results suggest that most arsenic species occur in a very stable organic form in seaweed which gradually transform to inorganic As(V) by sulfuric acid treatment at higher temperature.

5 References

1. Kudo K, Suzuki N (1987) J. Radioanal. Chem. *26*: 327
2. Kudo K, Suzuki N (1984) Trends Anal. Chem. *3*: 20
3. Klas J, Tölgyessy J, Klehr EH (1974) Radiochem. Radioanal. Lett. *18*: 83
4. Kyrs M, Prikrylova K (1983) J. Radioanal. Chem. *79*: 103
5. Klas J, Tölgyessy J, Lesny J (1977) Radiochem. Radioanal. Lett. *31*: 171
6. Klas J (1981) Radiochem. Radioanal. Lett. *47*: 355
7. Lesny J, Tölgyessy J, Rohon O, Stefanec J, Klas J, Chacharkav MP (1980) Radiochem. Radioanal. Lett. *42*: 9
8. Lesny J, Tölgyessy J, Rohon O, Zacharová Z (1981) Radiochem. Radioanal. Lett. *47*: 293
9. Kyrs M (1965) Anal. Chim. Acta *33*: 245
10. Kyrs M, Halova J (1979) J. Radioanal. Chem. *52*: 53
11. Kyrs M, Halova J (1983) J. Radioanal. Chem. *78*: 29
12. Downey DM, Huston GC (1983) Anal. Lett. *16*: 1469
13. Shikano K, Kudo K (1983) J. Radioanal. Chem. *78*: 71; (1982) ibid. *74*: 73
14. Katoh M, Kudo K (1984) J. Radioanal. Nucl. Chem. *84*: 277
15. Akolzina LD, Alimarin IP, Bilmovich GN, Churkina NN (1980) J. Radioanal. Chem. *57*: 279
16. Yakovlev YnV, Kolotov VP (1981) Zh. Anal. Khim. *36*: 629
17. Mitchell JM (1984) Mikrochim. Acta 243
18. Shigematsu T, Kudo K (1981) J. Radioanal. Chem. *67*: 25
19. Shigematsu T, Kudo K (1981) J. Radioanal. Chem. *67*: 307
20. Ruzicka J, Stary J (1961) Talanta *8*: 228
21. Ruzicka J, Stary J (1961) Talanta *8*: 535
22. Suzuki N (1958) Proc. 2nd. Conf. Radioisotope (Japan) 151
23. Kudo K, Suzuki N (1980) J. Radioanal. Chem. *59*: 605

24. Bilimovich GN (1985) J. Radioanal. Nucl. Chem. *88*: 171
25. Suzuki N, Imura H (1987) Bunseki (Analysis) 106
26. Bilimovich GN, Atrashkevich VV, Alimarin IP (1974) J. Anal. Chem. USSR *29*: 555
27. Briscoe GB, Dodson A (1967) Talanta *14*: 1051
28. Katoh M, Kudo K (1983) Bunseki Kagaku (Anal. Chem. Japan) *32*: 1
29. Katoh M, Kudo K (1983) J. Radioanal. Chem. *79*: 23
30. Katoh M, Kudo K (1985) J. Radioanal. Nucl. Chem. *90*: 277
31. Yuzawa M, Suzuki N (1981) J. Radioanal. Chem. *62*: 115
32. Suzuki N, Yoshida K, Imura H (1981) Anal. Chim. Acta *129*: 221
33. Suzuki N, Nakamura S, Imura H (1984) J. Radioanal. Nucl. Chem. *81*: 37
34. Suzuki N, Takahashi M, Imura H (1984) Anal. Chim. Acta *160*: 79
35. Suzuki N, Hanzawa K, Imura H (1986) J. Radioanal. Nucl. Chem. *97*: 81
36. Langer A (1950) Anal. Chem. *22*: 1288
37. Suzuki N (1959) Nippon Kagaku Kaishi (J. Chem. Soc. Jpn.) *80*: 373
38. Ikeda N, Takahashi Y (1977) J. Radioanal. Chem. *36*: 77
39. Rao VRS, Tataiah G (1976) Radiochem. Radioanal. Lett. *24*: 275
40. Rao VRS, Tataiah G (1976) Radiochem. Radioanal. Lett. *25*: 223
41. Rao VRS, Tataiah G (1983) Radiochem. Radioanal. Lett. *57*: 113
42. Acharya BV, Rao YK, Reddy GS, Rangamanner B, Rao VRS (1985) Radiochem. Radioanal. Lett. *94*: 357
43. Pareedulla K, Acharya BV, Hussain RC, Rangamannar B (1985) Radioisotopes *34*: 421
44. Polak HL (1971) J. Radioanal. Chem. *9*: 241
45. Polak HL, Groot HDe, Challa EE (1973) J. Radioanal. Chem. *13*: 319
46. Kambara T, Suzuki J, Yoshioka H, Nakamura T (1975) Radioisotopes *24*: 755
47. Kambara T, Suzuki J, Yoshioka H, Nakamura T (1975) Chem. Lett., 927
48. Kambara T, Suzuki J, Yoshida H, Ugai Y (1980) J. Radioanal. Chem. *59*: 315
49. Kambara T, Suzuki J, Yoshioka H, Watanabe Y (1980) J. Radioanal. Chem. *60*: 121
50. Kambara T, Suzuki J, Yoshioka H, Nakamura T (1980) Radioisotopes *29*: 590
51. Kanda Y, Suzuki N (1979) Radiochem. Radioanal. Lett. *37*: 183
52. Kanda Y, Suzuki N (1980) Anal. Chem. *52*: 1672
53. Imura H, Suzuki N (1981) Anal. Chim. Acta *126*: 199
54. Imura H, Suzuki N (1983) Anal. Chem. *55*: 1107
55. Kanda Y, Suzuki N (1979) Radiochem. Radioanal. Lett. *39*: 221
56. Suzuki N, Jitoh F, Imura H, Kanda Y (1987) Anal. Chim. Acta *193*: 239

Neutron Activation Analysis in Archaeological Chemistry

Garman Harbottle

Department of Chemistry, Brookhaven National Laboratory, Upton, NY 11973, USA

Table of Contents

Topics in Current Chemistry, Vol. 157
© Springer-Verlag Berlin Heidelberg 1990

Garman Harbottle

There is a long history of the application of chemical analysis to archaeological problems, extending to the last years of the 18th century. The nuclear-age technique of neutron activation analysis, permitting the simultaneous, sensitive, non-destructive estimation of many elements in an archaeological specimen, has found wide application. Important advances have been made, using this technique, in locating the origins of archaeological artifacts such as ceramics, metals, obsidian and semiprecious stones, among other articles of ancient ritual and commerce. In addition, the technique of neutron activation analysis has proved to be almost ideal in studies tracing the development of ancient technologies such as glass-making and smelting. In the future, the development of data banks of analyses of archaeological materials should provide an excellent new tool in studies of prehistory.

1 Introduction

The present paper is concerned with the varied uses of neutron activation analysis (NAA) in archaeology, but first a few words of history, outlining the interaction between archaeology and chemical analysis itself. If we may define "Archaeometry" as the application of the measuring techniques of the hard sciences to the remains left by ancient peoples, then it would seem that Archaeometry antedates Archaeology [1, 2, 3]. The first quantitative chemical analyses of archaeological artifacts were performed by Martin Klaproth and reported in a paper read July 9, 1795 [1]. The objects investigated were six Greek and nine Roman coins. Shortly thereafter Klaproth analyzed three specimens of Roman glass, one brilliant green, one bright red and one sapphire blue, which had been part of a glass mosaic in the ruins of the villa of Tiberius at Capri. His purpose was to discover the cause of the coloration, and to carry out the complete quantitative analysis of glass he was forced to invent the method itself, just as his earlier work had required him to invent a quantitative method for the analysis of copper-based metals. Thus the work of Klaproth brought important innovations in both the fields of archaeology and analytical chemistry.

In the first half of the 19th century there were a number of reports of chemical analyses of archaeological objects, some involving famous chemists like Sir Humphry Davy (ancient pigments from Rome and Pompeii) [4, 5], Berzelius [6], and Fresenius [7]. An especially significant study was that of Göbel, who was Professor of Chemistry at the University of Dorpat [8]. His work for the first time, in 1842, drew attention to the intimate connection between chemical analysis and the comprehension of the significance of archaeological remains: he suggested that the composition of brass objects excavated in the Russian Baltic provinces indicated contact with Rome, and went on to generalize that these copper-zinc alloys were in fact first produced in the Roman Empire. It is during this period of the 19th century that professional cooperation between chemist and archaeologist first became evident. The long title of Göbel's book "On the power of chemistry in identifying peoples of ancient times, or results of a chemical investigation of metallic antiquities, especially those occurring in the Baltic provinces, for the purpose of identifying the peoples from which they came" says it all. In 1853, A. H. Layard's archaeological publication, "Discoveries of the Ruins of Nineveh and Babylon" contained chemical analyses of Assyrian bronzes.

In 1853 and 1855 Wocel, an Austrian, published two papers on the analysis of specimens of ancient bronze which he had made, but in addition, summarized a number of earlier analyses [9, 10]. He took an important step forward when he argued that the relative dating, or seriation, of ancient metal specimens might be reflected in their chemical compositions. A number of additional examples of early archaeological chemistry are given in the excellent books and reviews of Earle R. Caley [1, 2, 3, 11, 12].

Fundamental to the chemical investigation of archaeological provenience is the concept that the composition of an object could give information as to its place of origin. In a paper published in 1977 [13] Weigand, Harbottle and Sayre termed this concept the "Provenience Postulate", namely "there exist differences in chemical composition between different natural sources that exceed, in some recognizable way, the differences observed within a given source". The germ of this idea is in the work of Fouqué [14] on the pottery found in the volcanic destruction layers at Thera (Santorin) and more clearly in the studies of Damour [15, 16, 17]. Damour's papers are concerned with the beautifully-finished hard-stone axes ("celts") which are found "in the Celtic monuments and among the wild tribes". These axes, which Damour studied both chemically and mineralogically, were cut from jade (jadeite and nephrite), chloromelanite, quartz, agate, jasper and fibrolite, and he drew the clear inference that their presence in excavatiɔns "to the bottoms of ancient tombs" far from the natural sources of such stone, could cast light on the migratory movements of peoples in prehistoric times. Damour also studied obsidian artifacts and sources in a systematic way. Finally, he was an early and emphatic advocate of what we would today call "interdisciplinary" investigation, stating that archaeologists must invoke the aid of geologists, zoologists, and paleontologists to give meaning to their discoveries. Mineralogy and chemistry must make known the characteristics and composition of the artifacts unearthed. It is astonishing to read these prophetic words of Damour written in the 1860's, ten years before the first archaeological excavations, at Olympia, to be conducted along modern scientific guidelines and with modern standards of reporting [2].

Finally, one should mention the work of Richards at Harvard in 1895 [18] on the analysis of Greek ceramics: "At the request of Mr. Edward Robinson, of the Boston Museum of Fine Arts, several analyses of ancient Athenian pottery were recently made at this laboratory . . . the interest of these analyses was mainly archaeological, turning upon the identity of the source of these remains with that of others found in other cities, . . .' and finally 'The variations in the relative amounts (of the elements determined) are singularly small, the range being not nearly so large as that given by Brougniart, in his "Traité des Arts Ceramiques"'. Hence, it is possible, that all of these specimens, which were picked up in the city of Athens itself, were the product of a local pottery." This observation again implies the Provenience Postulate, and equates consistency in elementary composition with identity of origin. It has been noted that, as a result of modern work, we can safely agree with Richards that Athenian pottery is indeed remarkably uniform in chemical composition.

In this brief historical sketch most of the chemical analyses mentioned were *via* classic wet chemical methods, requiring slow and painstaking laboratory work. In the first quarter of the 20th century, emission spectroscopy was applied: for example, Baudouin [19] found that large numbers of bronze and copper objects could be readily distinguished, while Weiss and coworkers [20] employed emission spectroscopy to check the more usual chemical analysis of a Greek bronze axe.

In 1936 Hevesy and Levi launched the field of neutron activation analysis [21, 22], today in routine use in laboratories worldwide, and for many systems, the methodology of choice. Had it not been for two additional developments, NAA would, however, have remained a laboratory curiosity, of little practical significance. The first of these was the discovery of the fission chain reaction leading to nuclear reactors producing useful neutron fluxes many crders of magnitude larger than those available before, and the second was the development of the sodium iodide (and afterward germanium) detector for efficient gamma ray measurement, coupled with the Wilkinson analog-to-digital converter and pulse height analyzer. These developments combined to make possible NAA as we use it today — fast, nondestructive, multielement, sensitive.

The first reports on the use of nuclear methods for analyzing archaeological materials were those of Ambrosino et al. [23, 24] and Emoto [25] dealing with coins, the detection of silver in a lead roof tile, and the proof of gold in a foil attached to a Korean sword. In 1954–56, at the suggestion of J. R. Oppenheimer, R. W. Dodson and E. V. Sayre began a series of studies on nuclear techniques in archaeology at Brookhaven Laboratory, which are still continued today. Among these early Brookhaven studies was a paper by Sayre [26] on the use of the lithium-drifted germanium detector for analysis of ancient glass; marking the first use of a Ge(Li) gamma detector in NAA.

2 The Uses of Neutron Activation Analysis in the Study of Archaeological Materials

2.1 Overall Advantages

The basic principles underlying NAA are simple, but they provide an analytical method that is different in many ways from other methods. While it is not necessary to discuss NAA in general [27] it is good to say why it is so useful in archaeological and fine-arts research, and at the same time to discuss some of the methodological problems of NAA research that have proved to be important.

NAA is non-destructive. That is one eventually recovers the sample and after the radioactivity has decayed sufficiently the specimen whose composition has been determined can be restored to a museum collection or further investigated. This non-destructive feature of NAA has been of great importance in some archaeological studies. In their investigations of Mesoamerican jade artifacts Bishop et al. [28] were able to determine 11 elements in whole artifacts (beads up to 15 mm diameter, weighing 9 grams) as well as in samples taken by drilling. By their ability to analyze whole objects non-destructively they obtained access to museum and privately-held specimens that could not otherwise have been studied. A second case is the author's unpublished collaborative research on the analysis of Mesoamerican bronze specimens [29]. Since these metal fragments had been removed from rare and unique archaeological finds, they were necessarily quite small, and non-destructive analysis was imperative. Despite sample sizes ranging down to 10 mg, it was possible to estimate concentrations of 10 elements, including indium to ppm and gold to ppb levels. A third example, also of the author, concerns the analysis of turquoise, another semi-precious stone used culturally, like jade, in preColumbian America [13]. Virtually all the hundreds of turquoise artifacts analyzed to date, comprising beads, amulets, worked chips or tesserae, and debitage, have been investigated non-destructively. In a recent set of analyses some twenty whole beads ranging from 30–200 mg could only be analyzed non-destructively because the Indian tribe on whose lands the artifacts had been found insisted on their reburial intact at the end of the archaeological study.

Although the specimen is retained following NAA, certain physical changes can occur as a result of the intense ionizing radiation and heavy-particle impacts that take place during neutron irradiation in the reactor. In particular, glassy and crystalline materials can change color. This problem is more serious for whole-object irradiations than for samples: in any case the color change is often reversible.

For many archaeological research projects involving NAA, destruction of the sample analyzed is not a serious matter. Pottery is usually drilled for removal of a specimen for analysis.

A second advantage, already mentioned, is that NAA is applicable to small samples. Also, it is by nature multielement, although the analytical sensitivity of each element (lower limit of detection) depends on nuclear quantities such as the reactor neutron flux, the capture cross-section, the branching ratio and energy of the particular characteristic gamma or X-ray observed and the half-life of the radioisotope. Thus the sensitivity can vary markedly across the Periodic Table. However, for purposes of provenience research, the multielement character is all-important, as will be seen in Section 2.4 below. It is usually possible by the mere inspection of a table of analytical data to say that the data was or was not obtained by neutron activation, because of the bias toward certain elements.

The dependence of the sensitivity upon half-life of the radioisotope concerned is most important in NAA, where one can often simply wait for interfering lines in the gamma-ray spectrum to die out in order to determine weaker lines of other radio-elements. For example, in the analysis of jade, which is typically 10% sodium, it was only possible to measure many of the other elements after the Na^{24} had decayed away, a period of two weeks [28].

Finally, the use of nuclear instrumentation in the NAA process has proved to be unusually well-suited to automation. Large numbers of samples may be simultaneously irradiated, their gamma-ray spectra determined by an apparatus consisting of automatic sample-changer coupled to a germanium detector with tape or disk readout, processed by mass computational methods. This results in very rapid multielement data accumulation, requiring computer storage and manipulation, as will be evident in the section below devoted to Provenience research.

The use of NAA in archaeological research suffers today from two major problems. The first is one of methodology: because the different laboratories practicing NAA tend to use different standards, results are not always transferable from one group of research workers to another. This point will be discussed below in Section 3. The second problem is economic. Analyses by commercial NAA run 70–100 dollars or more each, and this cost can inhibit widespread use of the method.

2.2 Technical Examinations

Neutron activation analysis has found ample use in research which has as its goal an understanding of ancient technology. That type of research has tended to fall into certain rather well-defined areas, which can now be discussed.

2.2.1 Glass

The analysis of ancient glass from a number of historic and prehistoric epochs has served to clarify our understanding of the development of glass technology. It will be remembered that among the first analyses of archaeological materials [11] were bits of colored Roman glass, and here it was the intention of Klaproth to understand the cause of the colors. The work of Sayre and colleagues [26, 30–34] first by emission spectroscopy and then including NAA, demonstrated that during specific historic periods such as the second millenium BC (Egyptian, Mycenaean, Sumerian), the Roman, and the Islamic, glass compositions tended to reflect a particular technological practice over large cultural areas. Also, in a number of cases the deliberate addition of particular compounds as colorants, decolorants or opacifiers could be documented [33]. Thus the changing practice of glassmaking could be followed through long sections of human history. These studies have been extended more recently into other historical periods [35-41] and geographic areas [42,43] with great success. Specific classes of glass, for example faience, have been investigated with a view toward understanding not only the provenience but also the technical methods of production, especially concerning color [44-47]. The famous Egyptian blue is a case in point [48].

2.2.2 Pigments

In the analysis of archaeological objects it often happens that the object investigated is itself a work of art, and the need to limit oneself to only small samples becomes evident. It will be remembered that Sir Humphry Davy [4] analyzed mural pigment specimens he himself had collected in Rome and Pompeii, and later [5] Roman specimens found in England [49]. In a more recent paper Ortega and Lee [50] found by NAA that the yellow pigments employed by ancient Mesoamericans at sites like Teotihuacan and Cholula could be characterized by neutron activation. A surprise was their finding that the former differed markedly from the latter in containing large concentrations of mercury. Thus, this pigment may once have contained cinnabar (HgS). In Fig. 1 is shown a ritual incense burner that was excavated at Teotihuacan, near Mexico City, although it is in the exact style of Monte Alban, some 600 km to the south and east [51]. Examination of the ceramic by neutron activation analysis at Brookhaven revealed mercury levels of 30–1500 ppm, indicating that the object was also at one time colored with cinnabar, even though none of the color was present today.

Although it is, strictly speaking, an application of NAA to the Fine Arts rather than

Fig. 1. Ritual incense burner in Monte Alban style excavated at Teotihuacan, near Mexico City, by J. Vidarte [51]

archaeology, it is worth mentioning in passing that the development of the technique of autoradiography of whole paintings is a further example of pigment analysis [52-55]. Sayre and colleagues working in this laboratory exposed whole oil paintings of Dutch masters and American 19th and early 20th century artists to neutron fluxes from a reactor: the paintings were then autoradiographed revealing the nature and spatial structure of the pigments employed. In particular, the characteristic brushwork of artists like Rembrandt, hidden beneath layers of paint, was revealed [55].

2.2.3 Metals

There have been an extraordinary number of analyses of archaeologically-significant pure copper and copper-based alloys (bronze of different types, brass) reported in the literature, but the great bulk of these before World War II were carried out by emission spectroscopy [56, 57]. X-ray fluorescence has also been used [58, 59]. However, as mentioned in Section 2.1 above, NAA is very advantageous for this type of work by virtue of small sample requirement and non-destructive methodology. This was especially important in the study of Friedman et al. [60] where analyses of Moche copper cups were reported: minute drilled samples showed trace element levels that permitted the conclusion that while some of the cups were fabricated from native copper, others were made from copper smelted from copper ores (see however [61, 62]). The question of native vs. smelted copper was also important in a study of artifacts of the "Copper Inuit" (Eskimos) carried out by Wayman et al., using NAA [63]. Certain trace elements were found to be diagnostic criteria for the two types of metal.

The work of the author on the analysis of Mesoamerican bronze has already been mentioned, but in several other cases the non-destructive or small scale of sampling came into play. Gilmore and Ottaway [64] discuss this point: in the work of Thompson and Lutz larger artifacts of bronze were also studied [65]. Finally, it should be mentioned that other methods involving neutron reactions have also been used in the study of ancient bronze. For example, Li Hu Ho has published the analysis of a 4000-year old Qi culture bronze mirror [66]. Similarly, Gilmore [67] has employed epithermal neutron activation. Glascock et al. have demonstrated the utility of prompt (capture) gamma rays in the non-destructive analysis of copper-based artifacts: the method is not however sensitive to trace elements [68].

Because of the desire to trace archaeological copper and copper-based alloys back to their copper ore sources, a substantial amount of research has been done on the behavior of trace elements in native copper and in the metal during the copper smelting process, using NAA. This topic is however more appropriately considered under the Provenience section below.

2.2.4 Bone

There have been several NAA studies of archaeological bone [69, 70]. A good bit of useful information can be derived from levels of the three elements calcium, barium and strontium. The strontium/barium levels appear to characterize the species, which can help in the identification of the bone's animal origin while the calcium/strontium data seem to be related to ancient diet. Chemical changes do appear to take place in bone under burial conditions [71-73], which may signal caution in interpreting elemental analyses.

2.3 Coins

Ancient coinage may be considered a special class of "archaeological metal" but it has been so extensively investigated that it deserves separate consideration. Here again we must pay tribute to Klaproth as a pioneer [1, 2] in that not only was he the first to analyze ancient coins, but in fact invented the method of quantitative metal analysis to do so, and this in the 18th century!

The work of Ambrosino and colleagues [23, 24] has already been mentioned. The early volumes of the journal Archaeometry (first published in 1958) also contain some exploratory research on the application of NAA to the problem of coin analysis. Kraay and Emeleus at Oxford [74–77] realized that NAA was a very practical solution to one of the requirements of ancient coin analysis, and that was, for the rare and most valuable specimens at least, that it be totally non-destructive. They also quickly discovered another benefit: because neutrons activated the whole coin, they provided a bulk analysis as opposed to x-ray fluorescence, which gave only the surface composition. They found that a Corinthian coin, thought to be silver, was in fact silver-plated over a copper core. For precise NAA studies, the self absorption of neutrons in solid gold coins requires a substantial correction to the flux. Self absorption in silver, less serious has also been investigated (see below).

The pioneering studies of Kraay and Emeleus were of silver coins from different parts of the Greek world at different time periods: they employed a sodium iodide gamma ray spectrometer and determined copper and gold quantitatively in large numbers of these coins. They had hoped to gain information on the possible sources of silver although they suspected that in some cases remelting of coins might obscure this information. They did however produce economic information on what seemed to be deliberate manipulation of metallic composition in several cases.

There have since been many other studies of coins using NAA [78]. One of these by Barrandon et al [79] illustrates very well the need for a close relationship between the neutron activation analyst and the numismatist. The study followed the decline in silver content of coins in a tightly-controlled temporal series, issued by the Emperor Constantine, during the years 313–340 AD. Economic history, including shifts in monetary policy, comparison between different mints, and progressive debasing of the currency can be seen. Counterfeit coins were also detected: because of the ease of making molds from authentic coins, and the high value of ancient coins to collectors, non-destructive authentication has been an important capability of NAA examination of coins. It is interesting that this study was performed using a Cf^{252} source of neutrons, which leaves almost no residual long-lived Ag^{110m}. Wyttenbach and Hermann [80] have dealt with some of the technical problems of NAA of silver coins, arising from self-shielding effects.

2.4 Provenience Research

2.4.1 Introduction

The "Provenience Postulate" has already been mentioned in the Introduction, whereby it was suggested that samples taken from either a natural source of materials, or artifacts made from material of a given source, tended to resemble each other in chemical composition more nearly than samples drawn from different sources. The idea, that locational information can be derived from chemical or mineralogical profile groupings and their intercomparison, hence can be obtained *post facto*, is by now firmly embedded in the literature of archaeological science. It is an idea with roots at least a century in the past [8, 15-17].

There is some justification for this idea: most importantly, it seems to work, to generate archaeologically-useful groupings and source relationships, and to help to distinguish classes of objects that might otherwise be difficult to distinguish. It has been checked with some natural sources.

It is helpful to distinguish between "processed" and "natural" archaeological materials. Examples of processed materials would be glass and bronze, while examples of natural materials would be obsidian and native copper. Although the mathematical techniques used to find and confirm natural groupings of specimens based on their elementary composition may be applied equally well to natural and processed materials, there is no *a priori* reason to expect the Provenience Postulate to operate in the latter category. And yet the literature of archaeological science contains many instances where man-made artifacts and materials are grouped compositionally, and these chemical characteristics used to derive cultural, and sometimes locational information. What apparently happens is that man, having devised a technology to produce a useful material (glass, smelted metals like copper, alloys like bronze) tends to stay with the same procedure, utilize ore from the same ore body, smelted in the same way, mixed with the same proportions of other metals, or, in the case of glass, sand for the silica content combined with relatively fixed proportions of alkalis, lime, wood-ashes, etc. [30, 31]. When colorants or opacifiers are added, these too tend to follow a culturally-prescribed proportion, drawing upon traditional sources. Thus random compositions tend not to occur, and a kind of chemical grouping based on a combination of source limitations and cultural patterns can often be observed [37, 40, 81].

We find empirically that the range of archaeological materials which can be analyzed for their provenience content extends from the totally "natural" material which is used culturally just as it is found (examples would be native copper, flint, turquoise, jade, obsidian, some clays) through materials having some processing (levigated or tempered clays) to materials which are entirely processed or artificial (bronze, smelted copper or silver, glass, ceramics derived from mixtures of clays, faience). In each of these three types basically the same procedures of numerical taxonomy have been applied to the analytical data although it is not always clear why such methods should work in the 2nd and 3rd instances.

2.4.2 Numerical Taxonomy[1] in Provenience Research

Archaeologists traditionally classify ancient artifacts by observation and measurement of attributes of those artifacts; for example, the length, shape, material and fabrication technique of projectile points, the paste color and texture, temper, decoration, thickness and porosity of ceramic potsherds. Obviously, some attributes are parametric (thickness, porosity), others are not (color, material, decoration) [82].

---

[1] The term "pattern recognition" is also used in the provenience literature [85]. We prefer the term numerical taxonomy for the whole process of "grouping by numerical methods of taxonomic units into taxa on the basis of their character states" [86].

In this section I shall discuss some of the general considerations in the use of this type of descriptive data as input to mathematical methods which can produce groupings of objects (they need not all be artifacts) on the basis of similarity in their attributes, and then show how this input data may in fact be the quantitative chemical analytical values generated by NAA or other analytical techniques. This is the procedure called "Numerical Taxonomy" and one must draw heavily on the work of Sneath and Sokal [83] to explore, understand and use it.

The artifact attributes mentioned above, both numerical and non-numerical, that can be used as bases for a classificatory system would be termed "characters" by Sneath and Sokal [84]: "a character is a property or feature which varies from one kind of organism to another", referring to biological systems. The value of a particular character (red, brown, black color of potsherd surface, length in centimeters of a flint spear point) is termed a "character state" [84], which can thus be qualitative or quantitative. Quite clearly, the chemical analysis of an archaeological object measures the concentration level, the "state", of a series of "attributes" or "characters", the chemical elements.

In the search for groups or classes of these objects, based on the numerical taxonomy of their chemical compositions, one must subscribe first of all to the view of Sneath and Sokal that "natural" groups do exist and are meaningful: "we believe that natural classifications are of great usefulness because when the members of a group share many correlated[1] attributes, the 'implied information' or 'content of information' is high." [87].

Taxonomy recognizes two basic types of grouping procedures, the monothetic in which the defining set of features is unique, and each member of the group must have each feature to the degree stipulated, and the polythetic, where group members "are placed together that have the greatest number of shared character states, and no single state is either essential to group membership or is sufficient to make (an organism) a member of the group" [88]. In the development of mathematical procedures for multivariate group formation, where all the characters are present but in statistically-distributed states in each taxon (group member) it is clear that our "natural" groups are more nearly polythetic than monothetic in nature.

Although by implication we will be discussing the taxonomy of objects having multivariate numerical descriptions — i.e., describable as vectors mathematically, the non-numerical characters may also be used in classificatory schemes [89] and even "mixed" with the numerical characters [90].

When we turn to a consideration of the use of chemical analytical values in the establishment of archaeological groups, we immediately encounter the problem: which elements shall we use? Obviously we are biased toward those elements measurable by NAA, and might be tempted to limit ourselves to elements that can be determined to a certain arbitrary level of precision, say $\pm 1\%$ or $\pm 10\%$. Sneath and Sokal [91] raise the same question "(2) Are all characters of equal value and information in providing evidence on phenetic similarity?" If they are not, then clearly weighting procedures ought to be employed, i.e., if iron concentration is worth more than cobalt in establishing taxonomic groupings [92] it should be so weighted as input. Although it may develop that some elements are more useful than others in classification and

---

[1] "correlated" is not used here in the strict mathematical sense.

discrimination of certain sets of groups based on chemical similarity in general one must rigorously follow the first two of the neo-Adansonian principles enunciated by Sneath and Sokal [93]:

1. The greater the content of information in the taxa of a classification and the more characters on which it is based, the better a given classification will be.
2. *A priori*, every character is of equal weight in creating natural taxa.

If it is true in a particular case that some elements are more valuable than others in creating taxa, this information will fall out of the appropriate multivariate program automatically and need not be assumed at the outset.

The methodology of handling data distributed in multivariate hyperspace has been extensively discussed [94-99] and there are an abundance of multivariate statistical and taxonomic programs from which to choose. Typically these programs are arranged in "packages" in such a way that one type of input data can serve a number of multivariate computations [100-102]. The setting-up, operation and retrieval of multivariate data banks has not been so widely published, and will be treated in Section III below.

2.4.3 Distributions in Natural Compositional Taxa

The Provenience Postulate asserts that there are taxa of natural materials, including those studied in archaeology, that are recognizable by similarities in elementary composition occurring within each taxon. But the second principle of Sneath and Sokal (above) tells us that the characters — the elementary concentrations — ought to be taken at equal weight. The problem is that NAA can determine elements present at greatly varying concentrations: for example from elements like iron, sodium, potassium and calcium at percent levels down to elements like europium and lutecium at parts per million to parts per billion. Both the abundant and trace elements carry provenience or group-forming information and one would like to give their variation in specimens analyzed equal weight. Two procedures are used: standardization and conversion to log (concentration) values. Since $\log A - \log B = \log A/B$, by prior log transformation we give the effect that relative changes (i.e., from one sample to another) in the concentration of an element are given equal weight regardless of the absolute level of concentration at which the element is present. An additional argument for the log transformation is that in natural occurrences trace elements can be found in log normal distributions [94]. For narrow distributions there is little practical difference in using the normal or log normal distributions. The distribution of variates is often assumed to be normal or log normal in making calculations of probability of group membership.

2.4.4 Examples of Materials that Have Been Analyzed

2.4.4.1 Glass

The cultural determinants in glass composition, the deliberate choices characteristic of the Egyptian 2nd millennium, the Roman and Islamic epochs, have already been mentioned. But on another level, within a particular temporal/geographical band, chemical regularities which can be used for classification may be observed in glass compositions. For example, Christie et al. [37] applied multivariate methods to group

Roman glass found in Norway by chemical composition. A rather more extensive study has recently been reported by Frana et al., who employed NAA to determine up to 32 elements in ancient Bohemian glass [43]. They found that "there is a marked grouping of glass products of particular periods, i.e., the late Hallstatt to early La Tene period and late La Tene period (Celtic) glass, which differ from the samples of Roman glass analyzed by us and naturally, from modern products". One may also mention the NAA work of Sanderson and Hunter on 1st millennium [36] and Roman-Post Roman and Medieval Glass [35], and the faience studies of Aspinall, Warren and coworkers [45, 47].

2.4.4.2 Metals

In the case of metals, a very substantial amount of research has been devoted to obtaining provenience information through chemical analysis. In many of the cases cited here, the chemical analysis was predominantly by NAA.

Craddock, in a comprehensive paper on medieval and west African bronze analysis, has clearly pointed out the inherent dangers of interpretation [57] of analytical data for copper or copper-based alloys in terms of provenience. "Provenancing metal from its composition has always proved immensely difficult even for the ancient Near East and Europe where one can normally assume that the metal has been mined, smelted, fabricated, used and discarded within the same society. In the case of West Africa where the bulk of the metal came from a wide variety of undifferentiated sources which were far distant and technically superior, the task of interpreting the metal analyses becomes all the more fraught with difficulty. In copper provenancing studies generally three assumptions have to be made:
1) that the copper from one mine or area has a distinct but uniform composition . . .
2) that the trace element composition is unaffected or at least consistently varied by the smelting process . . .
3) that there is no intermixture of copper from different sources . . ."
He then examines some of the evidence regarding the validity of these assumptions. I might add parenthetically that in my own laboratory the analysis of a large number of specimens of turquoise, a copper phosphate mineral, which had been obtained from known mining areas, has demonstrated that Assumption (1) of Craddock's paper is true only in selected instances. Our analysis of other copper minerals like malachite also showed great trace element variance within a given source in several cases.

Despite these warnings of Craddock, which also apply to other mined and smelted metals like silver and iron, there have been serious attempts to glean locational information from analytical data. Berthoud, in his thesis research [103] and in a paper published with several collaborators [104], using plasma emission and spark source mass spectroscopy, analyzed the multivariate compositional data of copper ores from more than 25 copper mines in Iran, that would have been important in early (4th and 3rd millennium) metallurgy. Their feeling was that the Craddock Assumption 1) was satisfied well enough and furthermore that it was possible to trace certain 4th millennium objects from Susa to a native copper source at Talmessi. Of course, with native copper, Craddock Assumptions 2) and 3) were not tested.

Friedman et al. [105–106] along with Fields et al. [107], using NAA made a statistical study of the distributions of 6 elements Ag, Fe, Sb, Co, Hg and Sc in copper obtained from nearly 400 copper ores including native copper. They focussed on the three major *types* of copper sources: I. Native copper, II. Oxidized ores such as CuO or CuCO$_3$, including the malachite and azurite minerals, III. Sulfide ores such as Cu$_2$S, chalcopyrite and the like. Their statistical model asked the question whether one could distinguish copper derived from a particular one of these three ore types on the basis of analysis for the specified six elements and answered with a statistically-likely "yes". Although they calculated the matrix of 15 correlation coefficients among the trace element pairs, they unfortunately considered copper from *all* ores of a given type rather than those from only a particular mining area, where correlations might have been stronger. Nonetheless, their work does call into question the validity of the oft-quoted assertion of Maddin et al. [61] that ". . . no adequate criteria exist for distinguishing artifacts of native copper from those of worked and recrystallized smelted copper of high purity". It seems likely that Maddin et al. were mainly referring to criteria of metallographic structure.

When one is dealing with native copper, which was used aboriginally by non-metallurgical societies, the problem of trace element alteration through smelting of course does not apply. There have been a number of studies on compositional characterization of the provenience of native copper [60, 63, 105−109]. The Archaeometry Laboratory of the University of Minnesota (Prof. G. Rapp) has analyzed native copper from known cources by NAA, determining up to 48 elements of which 16 are used in taxonomy. They have used a simple numerical coefficient of matching that, in most cases, correctly assigns analyzed copper to its rightful origin [110].

A number of those reports concerned with the analysis of copper ores and changes on smelting also give analytical data for native copper, as mentioned above. There exists, by now, a substantial literature on that topic and it probably deserves a review article on its own because of the prehistoric importance of this material.

Rapp has thus shown that the six trace elements determined in copper by Friedman's group in the 1960's and '70's can by the use of more refined (and more time-consuming) NAA techniques be greatly expanded [111], and since in general numerical taxonomy and group discrimination all improve, the greater the number of elements considered it is likely that future studies will employ larger numbers of trace elements in the analytical scheme.

Other metals have been analyzed for possible provenience assignment *via* NAA and compositional profile matching. The work of Gordus [112], and then Sayre and his colleagues Meyers and van Zelst [113−115] on Sasanian silver was a sustained, and very fruitful effort. First, methods had to be devised for the removal of almost invisible samples (ca. 500 µg): with the extraordinary value of these museum pieces and the long life of Ag^{110m} no extensive sampling or whole-object irradiation could be contemplated. During this research more than 500 samples of Sasanian objects and related coins were analyzed for up to 16 elements. It was found that the trace gold and iridium content of the silver tended to characterize the silver source, while the style and iconography of the objects tended to correlate with the trace element profile. Chronological information could also be deduced. Finally, one could decide that, in general, silver coinage was *not* remelted to make vases, bowls and plates. The whole study demonstrates the value of an in-depth approach under the best possible conditions of collaboration between radiochemist and art-historian [115].

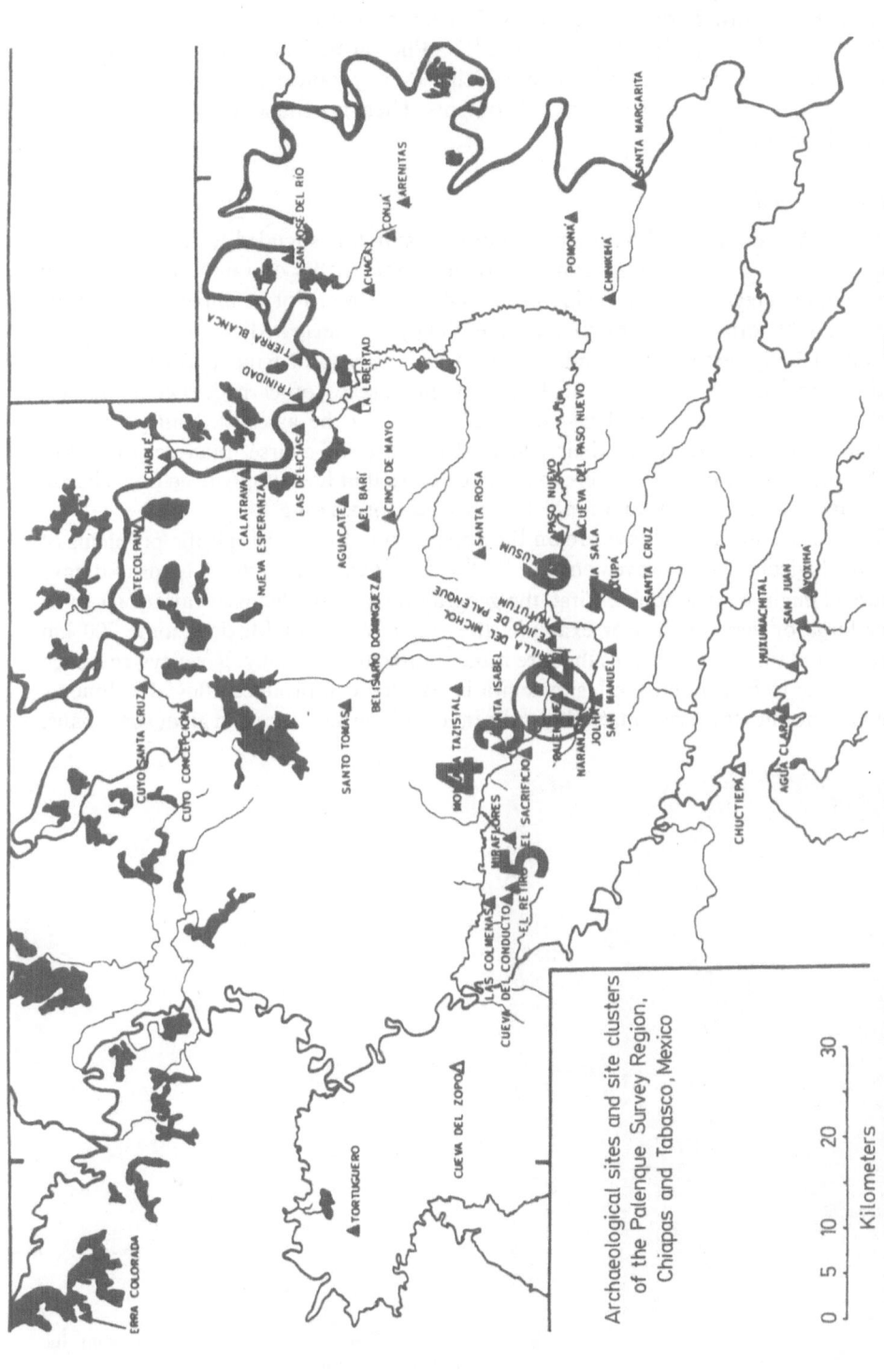

Fig. 2. Map of Mayan archaeological sites and clay resources of the Palenque region, Chiapas and Tabasco, Mexico [126]. Solid triangles are archaeological sites. Numbers refer to compositionally distinct clay resource zones

Other archaeological metal specimens that have been analyzed by NAA include the famous Moulsford Torc [116]: an Irish gold neck-ring weighing nearly a pound, and gold objects from Egypt [117] and Peru [118]. Photon activation analysis has been employed by Reimers et al. [119], avoiding the severe problem of self-shielding in the NAA of gold objects having large gold contents. There is a single study of Roman lead in the literature [120].

2.4.4.3 Ceramics

It should be noted that although provenience research has tended to regard archaeological ceramics as having the compositional characteristics of a natural material, in fact ceramic clays are often highly processed before use. For example, clays can be blended with other clays, as in the Nile Valley [121], levigated to provide a finer, smoother body texture [122] or mixed with a wide variety of temper to give strength and/or resistance to thermal shock. As in the case of other processed materials, archaeological ceramics analyzed by NAA can often be assembled into compositional groups that relate to their origin. There are, of course, many examples of archaeological ceramics that were made from clays that had simply been dug, shaped and fired without much additional technological processing.

There is a substantial literature on the application of NAA to specific problems of archaeological ceramic provenience [96,97,110,123−125] but a few general tends that have emerged might be mentioned. First, the geographic scale of chemical variation can be very large or very small. For example, the whole Valley of Mexico, some 200 km across, contains clay resources that are rather uniform chemically. River systems such as the Nile in Egypt and the Usumacinta in Mexico can produce alluvial sediments used in ceramics that are remarkably uniform over long distances and great time spans.

Fig. 3. Flanged Mayan Incensario from the Palenque region.

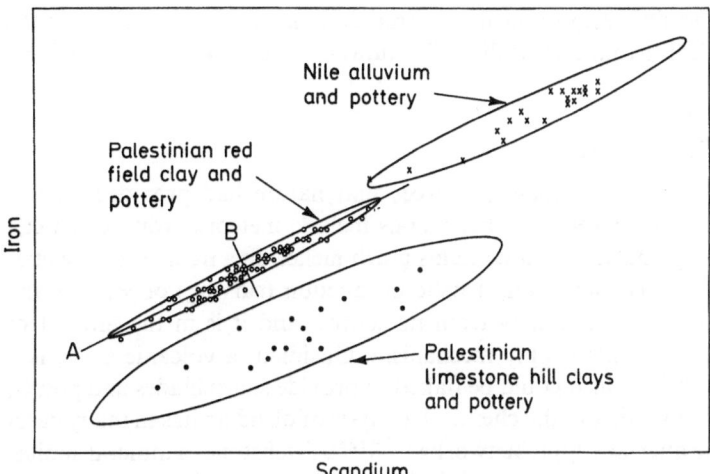

Fig. 4. Correlation of iron and scandium in three distinct groups of Middle Eastern clay and pottery specimens. In Palestinian red field clay and pottery group, A = axis of greatest variance, B = axis of least variance [94]

On the other hand, in more diverse geological settings, clay compositions can change drastically from point to point, sometimes in a space of a few hundred meters.

In Fig. 2, the map shows a region of Mexico in the states of Chiapas and Tabasco, containing the site of Palenque, (locations 1 and 2 in circle at center), one of the most important centers of the Maya civilization (First millenium AD). In the archaeological zone surrounding Palenque, the solid triangles indicate sites where fragments of pottery, including ritual incense burners (incensarios) have been found. A typical Mayan incensario is shown in Fig. 3. Analysis of these materials and an extensive sampling of clays by NAA showed that clay compositions from the regions numbered 1 to 7 could be distinguished from one another, and that the incensarios were all manufactured at Palenque itself (1 and 2). Clay sources 1 and 2 were only a few hundred meters apart [126].

Secondly, it is very commonly found in sets of analytical samples taken from a given clay resource that several pairs of elements show a high (r = 0.8 to 0.99) correlation coefficient [127, 128]. Frequently, iron and scandium are highly correlated [128, 129]. An example of this correlation of iron and scandium (which in the trivalent state have similar ionic radii) concentrations is given in Fig. 4 [94, 96]: the specimens analyzed are Nile alluvial sediments, the red clay that occupies much of the coastal plain of Palestine, and yellow clays derived from limestone formations in the hills around Jerusalem. Obviously, the location of points from a related group on a line having high correlation coefficient constitutes an additional important consideration in the numerical taxonomy of the group. It is surprising that even in highly-tempered pottery like the famous "thin orange" of Teotihuacan, many pairs of elements are found to be strongly correlated [130]. The existence of these correlation pairs, with the resultant non-sphericity of the compositional groups in hyperspace, on the one hand demands the use of Mahalanobis rather than Euclidean distance as the measure of similarity [94, 96, 131] but on the other provides a rather sharper mathematical tool for discriminating chemical taxa. It is certainly important that correlation matrices be examined in the course of any numerical taxonomic studies of natural materials like clay or obsidian (see below).

2.4.4.4 Stone and Mineral Products

Aboriginal man used culturally many resources that nature had provided: native copper has been mentioned already. Other metals include meteoric iron, which can be readily recognized by analysis, as it contains much nickel. In a number of societies stone was used for tools, construction, artistic production (carving) or jewelry and was transported considerable distances from its source, and it is in the interest of archaeology to reconstruct this supply mechanism. Obsidian, a volcanic glass, is a case in point. Of great utility, as it could be flaked to provide sharp blades and points, it was mined and transported, and the chemical analysis of obsidian has in many cases helped reconstruct the ancient supply networks [132-135]. Pitchstone, a mineral similar to obsidian but containing more water, is found in archaeological deposits in Britain and, like obsidian may be traced to its mining origin through NAA and x-ray fluorescence [136]. Additional discussion and references to obsidian will be found in Section 3:1 below. Flint and chert, also used for cutting tools have also been characterized analytically [137-142]: where obsidian tends to be chemically rather uniform in some deposits, flint and chert are much more heterogeneous. Nonetheless, they can in many cases be traced to source areas.

Sandstone, a material used in construction in Dynastic Egypt, has been the subject of interesting work by Heizer et al. [143,144]. The two (720 ton) Colossi of Memnon at Thebes are made of a quartzose sandstone ("quartzite"), not to be found in the vicinity.

Fig. 5. Fragment of a medieval limestone sculpture of an angel. Metropolitan Museum of Art (New York): the Cloisters [147]

The more northerly of the two is the famous "singing" Memnon: this statue was repaired by the Emperor Septimius Severus in Roman times. Heizer and coworkers analyzed the stone of the two statues and also that used in the repaired section, by NAA. The results indicated that the sandstone used in the original statues was quarried at Gebel el Ahmar, near Cairo, 676 km down the Nile, while the stone used in the Roman repair came from nearer quarries at Aswan and Edfu, upstream.

Limestone was important in antiquity not only as a building stone but also as a medium for sculpture: one thinks of the very widespread use of limestone for sculpture and hieroglyphic tablets among the Maya [145] as well as for construction.

A recent pair of papers by Sayre and colleagues [146–147] based on the work of Meyers and van Zelst [148] has shown that limestone quarried in France in medieval times and used for sculpture can readily be characterized by petrography and neutron activation analysis, and that even across the diverse, distinct layers of a quarry there are recognizable trace element patterns. One interesting result that has recently been published [147] concerns a fragment of Gothic relief carving of an angel, currently located at The Cloisters, a branch of the Metropolitan Museum of Art in New York. The sculpture (Fig. 5) has been assigned to the Abbey of Moutiers-Saint-Jean in the Côte d'Or region of France, on the basis of its sculptural style. But it was also thought possible that the sculpture had originated at the Abbey of Cluny (Saône-et-Loire, France). Groups of limestone sculptures known to have come from the two churches, now destroyed, were analyzed by NAA and the average chemical profiles established (Fig. 6.). Analysis of the angel fragment of unknown origin revealed close agreement with limestone used at Cluny but sharp disagreement with the compositional group related to Moutiers-Saint-Jean. These interesting results ought to stimulate research

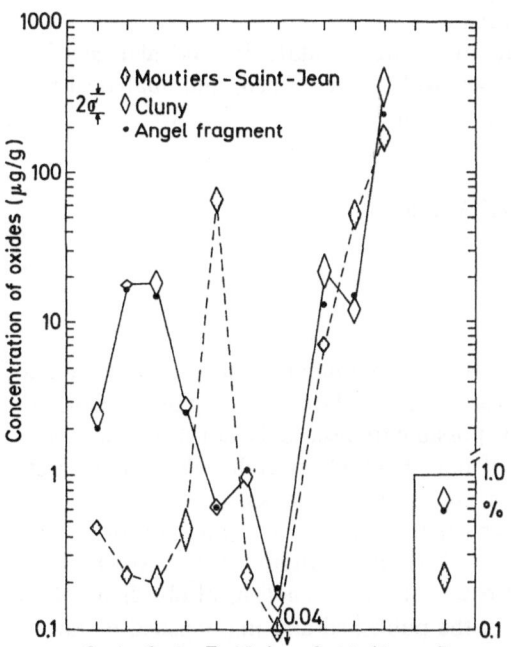

Fig. 6. Comparison of chemical composition of angel fragment (solid circles) with representative groups of limestone sculpture from Cluny (solid line) and Moutiers-Saint-Jean (dashed line). Vertical extension of diamonds = ±1 standard deviation [147]

in the use of NAA in characterizing archaeologically — used limestone as well, for example in the Maya world, and in Egypt.

Marble, also an important building and sculptural medium in the Greek world, has been widely studied petrographically [149] chemically [150-152] and through measurement of the isotopic ratios $^{13}C/^{12}C$ and $^{18}O/^{16}O$ [153-157]. The original studies using isotopic methods suggested that these alone might be able to give unequivocal provenience assignments. However it has been found that there are numerous overlapping source areas on the $\delta^{13}C$-$\delta^{18}O$ plots [155]. Likewise, trace element and petrographic data do not either, alone, give unequivocal assignments. Despite the fact that no perfect methodology exists, many valuable results have been obtained, and hybrid strategies for characterizing marble will no doubt be successful.

Another stone that has found wide cultural use as a carving medium in many early societies is steatite or soapstone, a very soft metamorphic rock related to chlorite and talc. In this laboratory, steatite (actually chlorite) from quarries near Tepe Yahya (Iran) was characterized by observation of the ratios of the relative intensities of basal-plane x-ray diffraction peaks [158] after it was found that NAA-determined trace element concentrations varied wildly within a given specimen. Another technique that has been used involves the determination by NAA of a number of lanthanide elements (La, Ce, Pr, Nd, Sm, Eu, Gd, Tb, Ho, Er, Tm, Yb and Lu) and the taxonomy of their abundances relative to each other — in other words, true "pattern recognition", when plotted as ratios to the levels of the same elements in a standard reference chondrite [159-160]. Although this technique found successful provenience application in several instances, in the recent work of Moffatt and Butler with steatite from the Shetland Oslands they found severe difficulties in obtaining uniform lanthanide patterns even within one quarry [161-162]. Any radiochemist wishing to embark on an archaeological research project involving steatite analysis should pay careful attention to the literature cited here, in advance.

Other stones which have received attention are sanukite [163] and alabaster [164]. The semiprecious stones jade and turquoise will be mentioned, and analytical provenience research on them referenced in Section 3 below.

3 The Storage, Use and Reuse of Data

3.1 Data Banks and Banking

We shall first survey the state of analytical data accumulation in several of the materials used culturally by ancient peoples and excavated or picked up in the course of archaeological field work. Then it will be possible to discuss the construction of data banks and the problems and promise of data banking in archaeological research, with several examples of successful applications of data banks.

Obsidian has, because of its heavy participation in ancient trade networks, been extensively studied by chemical analysis to establish its origin [165-173]. Much of this research has utilized NAA, which is almost ideal for the analysis of obsidian. Of the tens of thousands of obsidian analyses in the published and unpublished literature, many were generated in response to ad-hoc closed-end archaeological problems, but others involved broad-scale research considering cultures of large time depth, extending over wide geographic areas. In several instances, obsidian provenience determination has been the key to the formulation and testing of quantitative mathematical models for the relationships operating in ancient trade networks; a most valuable step forward since by implication these same models may equally well apply to the trade in other luxury or utilitarian resources of great value.

Analytical results for obsidian are spread over the records of a number of laboratories and the problems of standardization and data banking which I shall discuss with reference to ceramics apply equally well here. To my knowledge there has been no serious attempt to organize all obsidian data "under one roof" (i.e. with a common format for storage and more importantly a common or comparable basis for standardization). And yet it is clear that such an enterprise would yield substantial dividends. Obsidian is an especially favorable case for data banking because, like flint, chert and the semiprecious stones (jade, turquoise) it was used in its natural state, and there is thus the possibility of matching artifacts directly to sources. There is however a subtle problem that has emerged in obsidian analysis. It is usually implicitly assumed that the analysis of a few hand-picked specimens from a given source, mine or outcrop, will yield average values of the different elements with quite small standard deviations that reflect the average composition of the whole source and hence of the ancient artifacts taken from it. But the work of Bowman et al. [174] showed that this need not be so: that the concentrations of many pairs of elements in obsidian were highly correlated ($r \approx .99$) and furthermore that the individual analytical values lying along the bivariate correlation line actually gave mean values that had very large standard deviations. Thus crudely speaking it is the correlation lines rather than the simple analytical values that characterize this source. Work at Brookhaven has uncovered several other instances of similar behavior [175]: such correlations do not interfere with provenience assignment but in fact can make it much stronger. They cannot be

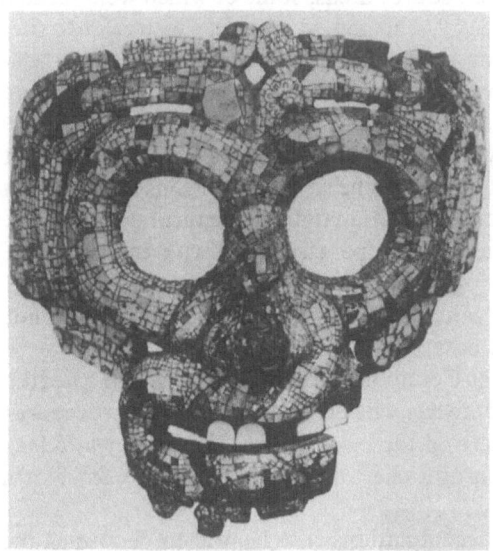

Fig. 7. Mexican Turquoise Mosaic Mask, Ethnographic Collections of the British Museum, London [179]

disregarded in general. Even where correlation does not play so large a role, there are but few cases where large obsidian outcrops have been adequately sampled.

With regard to data banking however, the existence of these cigar-shaped (in multivariate compositional space) as opposed to spherical groups of specimens coming from a given source strongly suggests that the data bank should contain as entries the analytical compositions of single individual specimens rather than average values, or average values with appended standard deviations. This would ensure that during the use of computer search programs (to be discussed below) there would be a better chance of making an initial identification "hit".

The situation with the semiprecious stones jade and turquoise is that data banks consist of the results of a relatively small number of investigations. Jade of Meso-america has been studied analytically [28] but not, to my knowledge, jade of China. Although the spreads of analytical values in jade from known individual sources are large useful characterizations have been achieved permitting much jade to be prov-enienced. An important result is the discovery of a group of Mesoamerican jade artifacts that can be analytically characterized but which cannot be related to any known source [28]; a finding that should stimulate further exploration.

In the case of turquoise the author and Weigand [13] are engaged in an open-ended project which seeks to analyze specimens taken from literally every existing and acces-sible turquoise source used aboriginally in North America. Early research by Sigleo likewise utilized NAA [176]. There is also another very extensive and detailed analytical study of turquoise recently published by Ruppert [177-178]. giving results for both North and South American source areas. In our work NAA is used, while Ruppert employed an automated electron microprobe. A data bank for turquoise exists, based on our work, at Brookhaven National Laboratory.

Turquoise was extensively used in several of the ancient Mesoamerican civilizations, the Toltec, Aztec and Maya, in the construction of mosaic. Fig. 7 is one of several turquoise mosaic masks in the Ethnographic Collections of the British Museum [179]. One of the purposes of the development of a turquoise analytical data bank is to relate the turquoise used in Mesoamerica to their source areas, some of which were located in the southwestern United States (Fig. 8) [13]. The named areas in Fig. 8 refer to sites important in mining, processing or trade in PreColumbian turquoise. Strong rela-tionships have been shown to exist between turquoise used in the Hohokam culture (7 in Fig. 8) and Chaco Canyon (site 6), by means of NAA.

Another pair of stones used culturally in their natural state are flint and chert. At first glance the lithic mine product would appear to be so heterogeneous as to suggest great difficulty in provenience attribution through analytical chemical profiles. But this is not the case. A number of studies in Europe associated with the names of Aspinall [137]; De Bruin [85]; and Sieveking [139-140] have shown that the chemical characterization of flint can in fact give valuable provenience information. I am not aware that there exists a European flint data bank.

In the United States the work of Ives at Columbia, Missouri [180-181] and Luedtke [141-142] come immediately to mind. Luedtke has written of the taxonomic procedures necessary to classify and provenience chert on the basis of NAA profiles, while Ives has been concerned with the compilation of a chert analytical data bank for North American resource areas.

In this discussion of analytical data accumulations it is helpful to choose as the

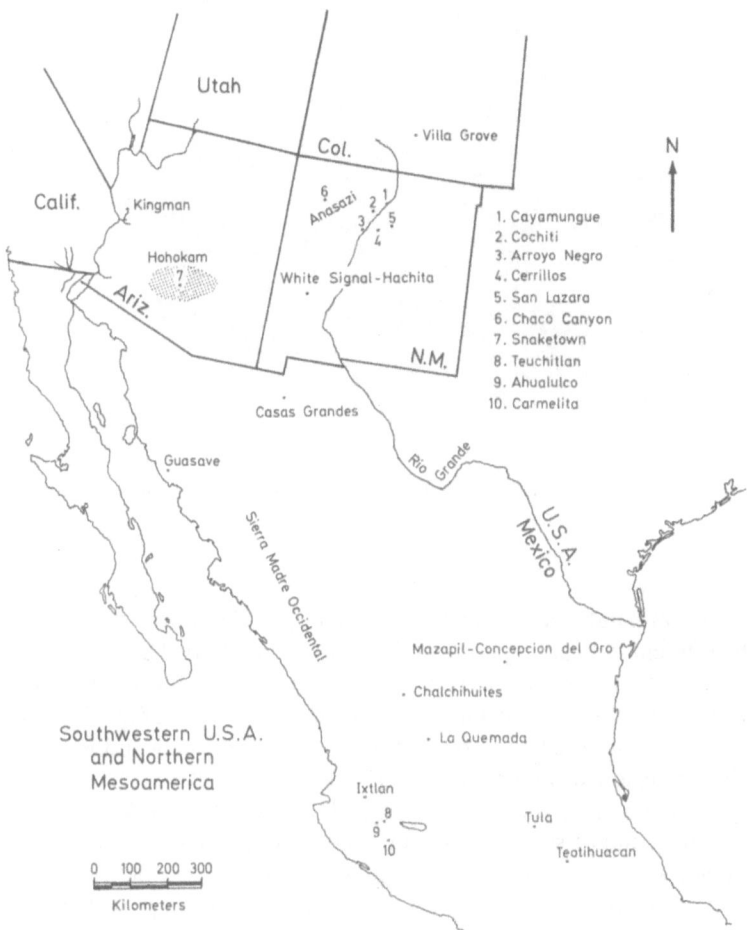

Fig. 8. American southwest and Mexico. Named sites and areas are locales important in the Pre-Columbian turquoise trade [13]

example of the ways in which data banks are set up and used the one material that perhaps has at once the greatest utility to archaeological research and the greatest extent of NAA involvement. That material is ceramics. It is not necessary to labor the great value of ceramic typologies and chronologies: they lie at the heart of much of our knowledge of archaeology from the Neolithic onwards. Ceramics from the Maya area of the New World and Greece in the Old were among the earliest archaeological materials systematically investigated by NAA [182–183], and today the method is widely used in a number of research projects.

There exist of the order of 50,000 analyses of archaeological ceramics, most all of which were obtained by neutron activation analysis. The recent development of inductively-coupled plasma emission (ICP) methods, which are frequently multi-element, relatively inexpensive and generally of quite acceptable sensitivity, precision and accuracy, promises that in the future that method will also make a significant contribution to archaeological analytical study.

Fig. 9. Six Preclassic period Mexican ceramic figurines from Tlatilco, Mexico City, plus a leg from a modern forgery

The ceramic analyses on hand were produced in a number of laboratories, mostly in connection with different *ad hoc* experimental schemes largely dedicated to the solution of specific, closed-end archaeological problems. In only a few cases were there studies aimed at a comprehensive knowledge of the chemical compositional profiles of all the ceramics, tempered and non-tempered, and all the clay and temper resources in a particular region of archaeological interest.

In the beginning of research in the analysis of archaeological ceramics, it was sufficient to make a visual comparison of the list of analytical values for the elements in a particular specimen with similar lists representing other specimens or with lists which were element-by-element averages of groups of specimens, the groupings having been formed by association of obviously-similar specimens. In Fig. 9 there are a group of figurines excavated from graves of the Preclassic period (800 BC) located in Tlatilco, now a suburb of Mexico City. The leg fragment numbered "8", at the right side, is from a modern forgery of the type sold to tourists and visitors. At Brookhaven, the figurines themselves, the leg, and clays from Tlatilco were analyzed by NAA. The results, (Table 1) show that both the ancient and modern specimens have the same compositions, making it highly likely that the figurines, like the fake leg, were a local product and not, as had been thought, imported [184]. An early refinement on this procedure involved the production of a chemical profile on a computer line-printer: The coordinates consisted of an X-axis proportional to the log of the concentration and a Y-axis with a fixed position for each element. Similarity in overall chemical composition between two specimens could then be found by superimposing the two computer plots upon one another over a strong light source [128]. A simple refinement involved the preparation of plots for comparison that represented group averages, with ± standard deviation limits, rather than individual specimens. Assignment of a specimen to a compositional group could then be made on the basis of these visual profile matches.

In the University of California (Berkeley) group of Perlman and Asaro matches of

Table 1. Comparison of the composition of six ancient figurines from Tlatilco with a modern Figurine (leg) and clays.

Oxides Determined	Average Concentrations in		
	Ancient Specimens	Modern Specimens	
Iron (Fe$_2$O$_3$)	6.75 ± 0.23	6.77 ± 0.70	
Sodium (Na$_2$O)	1.90 ± 0.10	2.02 ± 0.26	Percent
Potassium (K$_2$O)	0.99 ± 0.11	1.02 ± 0.11	
Manganese (MnO)	814 ± 188	1040 ± 660	
Barium(BaO)	535 ± 26	515 ± 7	
Chromium (Cr$_2$O$_3$)	128 ± 6	158 ± 30	
Cerium (CeO)	68.9 ± 5.4	67.6 ± 7.6	
Rubidium (Rb$_2$O)	66.1 ± 4.0	55.3 ± 7.2	
Lanthanum (La$_2$O$_3$)	36.9 ± 2.3	35.1 ± 0.6	
Cobalt (CoO)	27.3 ± 1.8	27.1 ± 5.9	Parts
Scandium (Sc$_2$O$_3$)	24.6 ± 0.6	24.8 ± 0.1	per
Thorium (ThO$_2$)	8.99 ± 0.34	8.42 ± 0.01	million
Hafnium (HfO$_2$)	6.37 ± 0.41	6.83 ± 0.17	
Cesium (Cs$_2$O)	4.60 ± 0.31	4.12 ± 0.25	
Europium (Eu$_2$O$_3$)	2.06 ± 0.17	2.14 ± 0.04	
Tantalum (Ta$_2$O$_5$)	1.06 ± 0.07	1.04 ± 0.11	
Terbium (Tb$_2$O$_3$)	0.91 ± 0.22	0.94 ± 0.04	
Antimony (Sb$_2$O$_3$)	0.53 ± 0.14	0.45 ± 0.16	

specimens to each other and to groups were carried out by inspection of the data, checking the assignments by probability considerations [185].

With the introduction of automated equipment (sample changers, tape readout) for NAA of large numbers of irradiated samples of archaeological ceramics in the early 1970's, and the resulting flood of analytical data relating to even one archaeological problem, it was realized that visual one-to-one comparisons of individual analytical profiles could not possibly keep pace. The number of comparisons that must be made goes up as $n(n-1)/2$, or essentially as n^2, where n is the number of samples analyzed. Also, visual comparison can give only a subjective measure of "similarity" and it would be much better to establish the criteria for chemical similarity on some more logical basis.

Because the Brookhaven group headed by Sayre was generating these massive quantities of data in the 1970's we were driven to seek more efficient methods for handling it. When the author became familiar with Sokal and Sneath's "Principles of Numerical Taxonomy" [83] and simultaneously D. L. Clarke's "Analytical Archaeology" [82] it became obvious that the characterization of archaeological ceramics through computerized multivariate statistical and numerical taxonomic methods was an urgent need, offering efficient ways of handling, manipulating and banking analytical data.

In the earliest phase of this effort, the analytical concentrations of a single analyzed specimen were encoded onto two IBM cards with eighty column format, of which six were given to identification numbers, one a letter denoting the substance (A = ceramics, W = obsidian, etc.) and one the number "1" or "2" for the first and second card. The remaining 72 columns per card were formatted as 18 groups, each of 4 columns, one group per element oxide. The two cards therefore carried the possibility of 36 elementary analytical concentrations per specimen. Each element was encoded as the digits, nnnm, where nnn were significant figures and m set the decimal place.

Thus the concentration range for each element ran from 99.9% (encoded as 9999) to 1 part in a trillion (encoded as 0010). Through the important part of this range, three significant figures were retained.

For many years the pairs of IBM cards punched in this way formed the data bank and the input to all the statistical and taxonomic programs of the Brookhaven group. In the 1980's however, Sayre developed three more formats for data storage, affording increasing opportunity for condensing the storage area requirement, and more adaptible to disk and tape storage than the IBM card images. At present six formats are available, and data in any of the six formats can be translated into any of the other five by the program FMTCHANGE, written by Sayre and now available in VAX version [186].

The data formats available are:

1. The two IBM card image format as described above.
2. 188-byte format with six-character identifaction, concentration data encoded as log values.
3. Same, but with seven character ID.
4. Arbitrary user provided format. Here, the program permits the user to specify the format, and this is of help in translating from one data bank to another.
5. Special format for editing. This format decodes the packed concentration data for 36 elements and presents it in a form which may readily be read, understood and edited at a terminal.
6. Condensed format. This is now the standard format for all storage at Brookhaven and at the Smithsonian Institution's SARCAR unit [186]. The data are stored in binary format, resulting in great compression. A single side of an 8" floppy disk will store the 36-element concentration data for each of 5800 specimens, while a single RLO2 hard disk would easily store all the chemical analytical data that have ever been compiled on archaeological materials.

There are at present a wide variety of multivariate statistical, taxonomic and other programs that accept as input the condensed (type 6) data, or, through FMTCHANGE, any of the other types including the user's own. There are several papers in which these Brookhaven-Smithsonian programs are described [94, 96, 186] while other workers in the field have utilized "package" statistical programs like BMD [102], SPSS [100], CLUSTAN [101], etc. which one need not review here. Instead, I would like to point out the utility of several additional computer programs developed by Sayre and Bishop at Brookhaven and at the Smithsonian [186] which are concerned purely with data bank operations. All of these programs also accept as input the condensed format (type 6) data mentioned above. The programs are:

1) SELECT which permits one to remove individual specimen condensed format data from a master file, to create a new subfile.
2) DELETE allows the deletion of data from a condensed file.
3) MERGE combines the data in two condensed files to form a third condensed file. In the event that analytical data for the same specimen and same element are found in both the input files, the program gives the operator a choice of incorporating either or neither value, or the average, to enter the output file. These three are

basic programs that are needed to manipulate, combine and split binary-packed files.

Finally, there are two programs for searching the entire data bank for specimens whose chemical compositional profiles agree with particular input criteria.

4) BSEARCH was written by Bishop in 1983 (in the VAX version XSEARCH). Here the analytical profile of a given sample is given to the computer. Also given are a string of element symbols which are the elements whose concentrations are to be used in calculating the mean Euclidean distance. Finally, a maximum mean Euclidean distance is entered. The computer than calculates the mean Euclidean distance [89,96] between the sample and, in turn, every other sample in the data bank, noting those whose distances are equal to or less than the specified distance. Thus the program locates the specimens chemically similar to the given specimen. The VAX can search a bank of 5000 specimens in about 30 seconds.

Sneath and Sokal, in "Numerical Taxonomy" [83] give among their fundamental principles that the more characters (i.e. chemical concentrations here) on which a taxon (natural grouping) is based, the better a given classification will be. In using any search programs ("pattern recognition" searches) like XSEARCH, the implication is that the more chemical elements ("characters") on the basis of which the search is conducted, for a given mean Euclidean distance, the more restrictive the identification of the target as "chemically similar" will be. On the other hand the high dimensionality of the multivariate spaces employed in the numerical taxonomy of compositional profiles leads to the near-equality, in terms of distance from the centroid, of the members of a chemical group [187]. Note that the hypervolume incremented by an increase in r, the Euclidean distance goes as r^n where n is the dimensionality. If n is large, let us say 15 or 20, this is a very powerful function, and the hypervolumes increase very steeply with unit change in r. For example, in 20-dimensional space (20 chemical elements chosen for the multivariate classification) an increase of only 15% in Euclidean distance results in a 16-fold increase in hypervolume. In such a space, Sneath and Sokal point out [187], if one has a roughly spherical cluster of points around a centroid, as the distance from the centroid increases, at first (small values of r) there will be very few points, within the distance r. However, as the radius increases, there will be a rapid increase in the number of new points until the edge of the cluster is reached. At this point the number of new points should drop abruptly. This effect produces a peak in the plot of points per unit increse in radius vs radius: many points lie at approximately the same distance, and very few points close-in. This is known as "Heincke's Law" [187]. With still larger increases in r one begins to pick up the outlying members of other groups in hyperspace, and in using XSEARCH it is a matter of trial-and-error to establish the best working conditions. These difficulties are partially compensated by the use of mean Euclidean distance as a dissimilarity measure.

5) ADSEARCH (XADSEARCH in the VAX version), written by Sayre and modified by Sayre and Bishop [186], operates in a different way. Fundamentally, the first portion of the program is based upon the full multivariate statistical program ADCORR [94]. ADCORR takes as input the standard (condensed, type 6) data file of up to 300 specimens, and up to 15 chemical elements per specimen. It then calculates a variance-covariance matrix for this "core group", and extracts the

eigenvalues and eigenvectors (the characteristic vectors). The correlation matrix can also be calculated. In ADCORR the probabilities of group membership can be calculated for each member of the core group and then, of each member of a comparison group of specimens in turn. In ADSEARCH the same basic steps are taken, but instead of calculating the probability of core-group membership of non-core group members, the whole data bank file is considered and every member of the data bank is taken as a potential core-group member. The printout then lists all the probabilities of core-group membership (multivariate normal distribution of chemical element concentrations in the core group is assumed), flagging those samples whose probability exceeds a preset value, say 5% or 10%.

Since both ADCORR and ADSEARCH involve matrix algebra, there must be a concentration value for each element for each specimen; i.e. the input data matrix must be complete. Both programs have several options for interpolating synthesized missing data, a problem of very wide occurrence in multivariate statistics. In both cases, the core group must have at least one more specimen than the number of variates (elements) considered, and ideally, should have several times as many.

3.2 Provenience Data: The Question of Intercomparability

Again I shall discuss this important question in terms of archaeological ceramics although, as stated above, the same problem of intercomparison applies to any other material as well. Until about the mid-1970's most analytical studies of ceramic provenience were self-contained — that is, any data accumulation was of an ad-hoc nature, and related only to the specific question in hand. However, the situation changed to one in which a number of laboratories began to generate large volumes of overlapping (geographically speaking) analytical data, and it was soon evident that such data could find valuable use outside the parent research group. It is obviously crucial to know whether data is in fact interchangeable among laboratories.

This problem, of analytical data quality control, is a general one in NAA practice. The most widespread NAA procedure today is to bombard the unknown specimens with neutrons in the presence of a standard or standards, whose elementary composition is assumed to be known. The author, in a study published in 1982, has examined the role of standardization in the NAA of archaeological ceramics and the requirements that must be met to permit the interchange of analytical data among laboratories [188]. On the purely technical side, what is necessary is that the standard used in the analysis of one group of specimens in one laboratory (or group of laboratories) be very carefully and precisely analyzed by the other laboratory using their customary standardization method. The author and C. J. Yeh [189] have published an intercomparison of this type deriving numerical values for the transformation of data between the two leading systems of ceramic standardization: the Brookhaven and Berkeley systems. One of the tasks of the SARCAR [186] unit of the Smithsonian Institution will be to organize data coming from a number of laboratories into a common standardization system so as to create a universal data bank, if that can possibly be achieved. Bishop, at SARCAR, has written a program XCONSTD, which converts data from one standardization base to another.

The need for interchangeability is also pressing in the case of obsidian data accumulations, where similar steps must be taken to promote exchange.

3.3 The Utilization of Data Banks

Analytical, provenience data banking is already of great, and will in future be of greater value to archaeological science. A few examples can be offered. The Archaeometry Laboratory at the University of Minnesota can now assign native copper, a most important trade commodity in early societies of North American Indians, to specific mining areas, with a high degree of reliability, based on NAA for some 16 elements [110]. Obsidian found in archaeological contexts almost anywhere in Mesoamerica can now be traced to its origin in one of a large number of Mexican and Guatemalan mines. Ancient trade routes in turquoise are beginning to be documented in the American Southwest: for example turquoise beads found in Hohokam contexts near Tucson Arizona have been found to match turquoise processing debitage at Chaco Canyon, far to the north [190]. Through the use of a data bank search program sherds of apparent Myceanaean style found far inland in an Iron-age cave in Jordan were shown to be of authentically Mycenaean origin [191]: on the other hand, "Greek" black-glazed sherds excavated near Marseilles France were a local imitation of that well-known export ware [192]. Small amphora sherds from the Greek period excavated at Carthage were demonstrated by data bank search to have come from Marseilles [193]. Many other examples could be adduced, and the constant growth of existing data banks augurs well for the future.

4 Conclusion

The archaeologist explores the hidden record of the earlier cultures of mankind, a record which in many instances gives us the only data we will have from which to reconstruct the religious, social and economic aspects of those long-gone cultural periods. A number of modern scientific techniques have been pressed into service to help interpret the archaeological record and among these techniques the nuclear sciences have played a leading role. Chemical analysis is one of the most important, permitting the archaeologist to identify materials used aboriginally, document processes of discovery and improvement in early technology, and establish the provenience of artifacts and materials that were the motivation of highly complex and extensive ancient trade routes.

Neutron activation analysis has proven to be a convenient way of performing the chemical analysis of archaeologically-excavated artifacts and materials. It is fast and does not require tedious laboratory operations. It is multielement, sensitive, and if need be, can be made entirely non-destructive. Neutron activation analysis in its instrumental form, i.e. that involving no chemical separation, is ideally suited to automation and conveniently takes the first step in data flow patterns that are appropriate for many taxonomic and statistical operations.

The future will doubtless see improvements in the practice of NAA in general, but in connection with archaeological science the greatest change will be the filling, interchange and widespread use of data banks based on compilations of analytical data. Since provenience-oriented data banks deal with materials (obsidian, ceramics, metals, semiprecious stones, building materials and sculptural media) that participated in trade networks, the analytical data is certain to be of interest to a rather broad group of archaeologists. It is to meet the needs of the whole archaeological community that archaeological chemistry must now turn.

Garman Harbottle

5 Acknowledgement

This research was carried out at Brookhaven National Laboratory under contract DE-AC02-76CG00016 with the U. S. Department of Energy.

6 References

1. Caley ER (1949) J. Chem. Educ. *26*: 242
2. Caley ER (1951) ibid. *28*: 64
3. Caley ER (1967) ibid. *44*: 120
4. Davy H (1815) Phil. Trans. *105*: 97
5. Davy H (1817) Archaeologia *18*: 222
6. Berzelius J (1836–7) Ann. Nord. Oldkundighed p. 104.
7. Fresenius R (1845) Ann. *53*: 136
8. Göbel F (1842) Ueber den Einfluss der Chemie auf die Ermittelung der Völker der Vorzeit oder Resultate der chemischen Untersuchung metallischer Altertümer insbesondere der in den Ostseegouvernements Volkommenden, Behufs der Ermittelung der Völker, von welchen sie abstammen. Erlangen
9. Wocel JE (1853) Sitz. ber. Akad. Wiss. Wien, Phil. hist. Klasse *11*: 716
10. Wocel JE (1855) ibid. *16*: 169
11. Caley ER (1962) Analysis of Ancient Glasses 1790–1957, Corning NY, Corning Museum of Glass
12. Caley ER (1964) Analysis of Ancient Metals, Oxford, Pergamon Press
13. Weigand PC, Harbottle G, Sayre EV (1977) Turquoise Sources and Source Analysis: Mesoamerica and the Southwestern U.S.A., in: Exchange Systems in Prehistory (eds. Earle TK, Ericson JE) p. 15, New York, Academic Press
14. Fouqué M (1869) Revue des Deux Mondes *83*: 923
15. Damour MA (1864) Comptes rendus *61*: 313
16. Damour MA (1864) ibid. *61*: 357
17. Damour MA (1866) ibid. *63*: 1038
18. Richards TW (1895) J. Am. Chem. Soc. *17*: 152
19. Baudouin M (1921) Comptes rendus *173*: 862
20. Weiss H, Dandurand F, Dureuil E (1923) Bull. Soc. chim. *33*: 439
21. Hevesy G, Levi H (1936) Kgl. Dansk Vidensk. Selskab., Mat.-Fys. Medd. *14*: 3
22. Hevesy G, Levi H (1938) ibid. *15*: 14
23. Ambrosino G, Pindrus P (1953) Rev. Metallurgy *50*: 136
24. Ambrosino G, Weill AP (1956) Bull. Lab. Musée du Louvre *1*: 53
25. Emoto Y (1957) Sci. Papers Japan, Antiques *13*: 37
26. Sayre EV (1966) Refinement in Methods of Neutron Activation Analysis of Ancient Glass Objects through the Use of Lithium Drifted Germanium Diode Counters, in: Comptes rendus 7th International Congress on Glass, Brussels, 1965, Paper 220, New York, Gordon and Breach
27. The best collection of papers devoted to the various aspects of modern NAA is to be found in the series of Proceedings of the Symposia, "Modern Trends in Activation Analysis". The 7th Symp. was June 23–27, 1986, Copenhagen. The Proceedings are in press, J. Radioanalyt. Nucl. Chem.
28. Bishop RL, Sayre EV, van Zelst L (1985) Characterization of Mesoamerican Jade, in: Application of Science in Examination of Works of Art, (eds. England PA, van Zelst L) p. 151, Boston, Museum of Fine Arts
29. Harbottle G, Hosler D, Lechtman H: unpublished thesis research of Hosler D
30. Sayre EV (1967) Summary of the Brookhaven Program of Analysis of Ancient Glass, in: Application of Science in Examination of Works of Art, p. 145, Boston, Museum of Fine Arts
31. Sayre EV, Smith RW (1961) Science *133*: 1824
32. Sayre EV, Smith RW (1974) Analytical Studies of Ancient Egyptian Glass, in: Recent Advances in Science and Technology of Materials, (ed. Bishay A) Vol. 3 p. 47, New York, Plenum Press
33. Sayre EV (1963) The Intentional Use of Antimony and Manganese in Ancient Glasses, in:

86

Advances in Glass Technology, Part 2, (eds. Matson FR, Rindone GE) p. 263, New York, Plenum Press

34. Olin JS, Thompson BA, Sayre EV (1972) Dev. Applied Spectroscopy *10*: 33
35. Sanderson DCW, Hunter JR (1981) Revue d'Archaeometrie *Suppl. Vol. III*: 255
36. Sanderson DCW, Hunter JR (1982) PACT *7:2*: 401
37. Christie OHJ, Brenna JA, Straume E (1979) Archaeometry *21*: 233
38. Davison CC (1972) Glass Beads in African Archaeology, PhD Dissertation, Berkeley, Univ. of California
39. Kuleff I, Djingova R, Penev I (1985) Glastech. Ber. *58*: 232
40. Kuleff I, Djingova R, Djingov G (1985) Archaeometry *27*: 185
41. Velde B, Gendron C (1980) ibid. *22*: 183
42. Frana J, Mastalka A (1984) Pamatky archaeologické *LXXV*: 458
43. Frana J, Mastalka A, Venclova N (1987) Archaeometry *29*: 69
44. Tite MS (1987) ibid. *29*: 21
45. Aspinall A, Warren SE, Crummett JG, Newton RG (1972) ibid. *14*: 27
46. Newton RG, Renfrew C (1970) Antiquity *44*: 199
47. Aspinall A, Warren SE (1976) The Provenance of British Faience Beads: a study using neutron activation analysis, in: Applicazione dei metodi nucleari nel campo delle opere d'arte, p. 145, Rome, Accademia Nazionale dei Lincei
48. Tite MS, Bimson B, Cowell MR (in press) The Technology of Egyptian Blue, in: Early vitreous materials (eds. Bunson M, Freestone IC) Occasional paper No. 56, London, British Museum
49. Farnsworth M (1951) J. Chem. Educ. *28*: 72
50. Ortega RF, Lee BK (1970) Archaeometry *12*: 197
51. Millon R (1973) The Teotihuacan Map, Austin, University of Texas Press
52. Sayre EV (1966) Revelation of Internal Structure of Paintings through Neutron Activation Autoradiography, in: Proceedings First Internat. Conf. on Forensic Activation Analysis, p. 119, San Diego
53. Lechtman HN (1966) Neutron Activation Autoradiography of Oil Paintings, MA Dissertation, New York, New York University
54. Sayre EV, Lechtman HN (1968) Studies in Conservation *13*. 161
55. Meyers P, Cotter MJ, van Zelst L, Sayre EV (1982) in: Art and Autoradiography (ed. Wasserman R) p. 105, New York, Metropolitan Museum of Art
56. Craddock PT (1985) Three Thousand Years of Copper Alloys: From the Bronze Age to the Industrial Revolution, in: Application of Science in Examination of Works of Art, (eds. England PA, van Zelst L) p. 59, Boston, Museum of Fine Arts
57. Craddock PT (1985) Archaeometry *27*: 17
58. Michel HF, Asaro F (1979) ibid. *21*: 3
59. Hedges REM (1979) ibid. *21*: 21
60. Friedman A, Olsen E, Bird JB (1972) American Antiquity *37*: 254
61. Maddin R, Wheeler TS, Muhly JD (1980) J. Archaeo. Science *7*: 211
62. Tylcote RF (1970) Antiquity *46*: 19
63. Wayman ML, Smith RR, Hickey CG, Duke MJM (1985) J. Archaeo. Science *12*: 367
64. Gilmore G, Ottaway BS (1980) ibid. *7*: 241
65. Thompson BA, Lutz GJ (1972) Radiochem. Radioanalyt. Lett. *9*: 343
66. Hu Ho Li (1985) Archaeometry *27*: 53
67. Gilmore GR (1976) Analysis of Ancient Copper Alloys using Epithermal Activation Techniques, in: Proceedings Fourth Internat. Conf. Modern Trends in Activation Analysis, p. 1187
68. Glascock MD, Spalding TG, Biers JC, Cornman MF (1984) Archaeometry *26*: 96
69. Wessen G, Ruddy FH, Gustafson CE, Irwin H (1977) ibid. *19*: 200
70. ibid. (1978) Trace Element Analysis in the Characterization of Archaeological Bone, in: Archaeological Chemistry II (ed. Carter GF) p. 99, Washington, Amer. Chem. Soc.
71. Edward J (1987) Post-mortem Changes in Archaeological Bone Composition and Their Effect on Paleodietary Reconstruction, PhD Dissertation, Columbia, Univ. of Missouri
72. Rottlander R (1976) J. Archaeo. Science *3*: 83
73. Rottlander R (1976) Zentralblatt für Geologie u. Paläontologie II, 5/6: 377
74. Kraay CM (1958) Archaeometry *1*: 1
75. Ibid. (1958) *1*: 21

76. Ibid. (1959) *2*: 1
77. Emeleus VM (1958) ibid. *1*: 6
78. The reader is referred to the work of Carter GF and a number of other specialists in this field, to be found in the pages of the publication Archaeometry. See Archaeometry *28* (appendix following p. 222) (1986) for a Cumulative Index.
79. Barrandon JN, Callu JP, Brenot C (1977) Archaeometry *19*: 172
80. Wyttenbach A, Hermann H (1966) ibid. *9*: 139
81. Ottaway BS (1979) Archaeophysika *10*: 597
82. Clarke DL (1968) Analytical Archaeology, London, Methuen and Co.
83. Sneath PHA, Sokal RR, (1973) Numerical Taxonomy, San Francisco, W. H. Freeman
84. Sneath PHA, Sokal RR (1973) ibid. p. 71
85. DeBruin M, Korthoven PJM, Bakels CC, Groen FCA (1972) Archaeometry *14*: 55
86. Sneath PHA, Sokal RR: op. cit. p. 4
87. Sneath PHA, Sokal RR: ibid. p. 25
88. Sneath PHA, Sokal RR: ibid. p. 21
89. Sneath PHA, Sokal RR: ibid., Chapt. 4
90. Philip G, Ottaway BS (1983) Archaeometry *25*: 119
91. Sneath PHA, Sokal RR: op. cit. p. 96
92. Sneath PHA, Sokal RR: ibid. pp. 109–113
93. Sneath PHA, Sokal RR: ibid. p. 5
94. Sayre EV: Brookhaven Procedures for Statistical Analysis of Multivariate Archaeometric Data, Brookhaven Laboratory Report BNL 21693
95. Sabloff JA, Bishop RL, Harbottle G, Rands RR, Sayre EV (1982) Analyses of Fine Paste Ceramics, in Excavations at Seibal, (ed. Willey GR) Memoirs of the Peabody Museum of Archaeology and Ethnology *15*(2): 266
96. Harbottle G (1976) Activation Analysis in Archaeology, in: Radiochemistry, Vol. 3. A Specialist Periodical Report, p. 33 (ed. Newton GWA) London, The Chemical Society
97. Harbottle G (1982) Chemical Characterization in Archaeology, in: Contexts for Prehistoric Exchange, (eds. Ericson JE, Earle TK) p. 13, New York, Academic Press
98. Sneath PHA, Sokal RR: op. cit., Chapters 4, 5 and 8
99. Everitt J (1977) Cluster Analysis, London, Heinemann Educational Books
100. Nie NH, Hull CH, Jenkins JG, Steinbrenner K, Bent DH (1975) SPSS Statistical Package for the Social Sciences, New York, McGraw-Hill
101. Wishart D (1978) CLUSTAN User Manual, Program Library Unit, Edinburgh, Edinburgh Univ.
102. Dixon WJ, Brown MB (1977) BMDP-77 Biomedical Computer Programs, Berkeley, Univ. of California
103. Berthoud T (1979) Etude par l'analyse de traces et la modelisation de la filiation entre minerai de cuivre et objects archaeolǫgiques du moyen-orient, thesis, Docteur es Sciences Physiques, Paris Université Pierre et Marie Curie
104. Berthoud T, Besenval R, Cesbron F, Cleuziou S, Pechoux M, Francaix J, Liszak-hours J (1979) Archaeophysika *10*: 68
105. Friedman AM, Conway M, Kastner M, Milsted J, Metta D, Fields PR, Olsen E (1966) Science *152*: 1504
106. Bowman R, Friedman AM, Lerner J, Milsted J (1975) Archaeometry *17*: 157
107. Fields PR, Milsted J, Hendrickson E, Ramette R (1971) in: Science and Archaeology, (ed. Brill RH) p. 131, Cambridge, MIT Press
108. Goad SI, Noakes J (1978) Prehistoric copper artifacts in the Eastern United States, in: Archaeological Chemistry II, (ed. Carter GF) p. 335, Washington, Amer. Chemical Society
109. Veakis E (1979) Archaeometric Study of Native Copper in Prehistoric North America, Ph.D. Dissertation, Stony Brook, Department of Anthropology, State Univ. of New York
110. Rapp Jr, G (1985) The Provenance of Artifactual Raw Materials, in: Archaeological Geology, (eds. Rapp Jr, G, Gifford JA) p. 353, New Haven, Yale University Press
111. Hölttä P, Rosenberg RJ (1986) Determination of the Elemental Composition of Copper and Bronze Objects by Neutron Activation Analysis, in: Abstracts of papers accepted at the 7th Internat. Conf., Modern Trends in Activation Analysis, p. 781, Copenhagen, RISO Laboratory

112. Gordus AA (1971) Rapid Nondestructive Activation Analysis of Silver in Coins, in: Science and Archaeology, (ed. Brill RH) p. 145, Cambridge, MIT Press
113. Meyers P, van Zelst L, Sayre EV (1973) Radioanal. Chem. *16*: 67
114. Meyers P, van Zelst L, Sayre EV (1974) Major and Trace Elements in Sasanian Silver, in: Archaeological Chemistry, (ed. Beck CW) p. 22, Washington, DC, Amer. Chemical Society
115. Harper PO, Meyers P (1981) Silver Vessels of the Sasanian Period, Vol. 1: Royal Imagery, New York, Metropolitan Museum of Art
116. Hall ET, Roberts G (1962) Archaeometry *5*: 28
117. Lefferts KC (1969) Bull. Metropolitan Museum of Art *28*: 61
118. Lechtman HN, Parsons LA, Young WJ (1975) in: Studies in Precolumbian Art and Archaeology No. 16, p. 7, Washington, D.C., Dumbarton Oaks
119. Reimers P, Lutz GJ, Segebade C (1977) Archaeometry *19*: 167
120. Wyttenbach A, Schubiger PA (1973) ibid. *15*: 199
121. Cockle H (1981) Jour. Roman Stud. *72*: 93
122. Cook RM (1960) Greek Painted Pottery, p. 242, London, Methuen
123. Rapp Jr G, Gifford JA (1985) Appendix: A Selective Bibliography, in: Archaeological Geology, (ed. Rapp Jr G, Gifford JA) p. 378, New Haven, Yale University Press
124. Riederer J (1981) Zum Gegenwärtigen Stand der naturwissenschaftlichen Untersuchung antiker Keramik, in: Studien zur altägyptischen Keramik, (ed. Arnold D) p. 193, Mainz
125. Sayre EV, Meyers P: Art Archaeo. Tech. Abstracts (1971) *8*: 115. The AATA routinely abstract ceramic provenience studies carried out by NAA and other means.
126. Bishop R, Rands RL, Harbottle G (1982) A Ceramic Compositional Interpretation of Incense-Burner Trade in the Palenque Area, Mexico, in: Nuclear and Chemical Dating Techniques, (ed. Currie LA) p. 411, Washington, D.C., Amer. Chem. Soc.
127. Harbottle G (1970) Archaeometry *12*: 23
128. Brooks D, Bieber Jr AM, Harbottle G, Sayre EV (1974) Biblical Studies through Activation Analysis of Ancient Pottery, in: Archaeological Chemistry, (ed. Beck CW) p. 48, Washington, D.C., Amer. Chem. Soc.
129. Turekian KK, Katz A, Chen L (1973) Limnology and Oceanography *18*: 240
130. Abascal Macias R (1974) Analysis por Activacion de Neutrones: una Aportacion para la Arqueologia Moderna, Thesis for Master of Archaeological Sciences Degree, Mexico, National Univ. of Mexico
131. Sneath PHA, Sokal RR: op. cit. p. 127 and p. 403
132. Hodder I (1974) World Archaeology *6*: 172
133. Renfrew C, Dixon J (1977) Obsidian in Western Asia: a Review, in: Studies in Economic and Social Archaeology, (eds. Longworth I, Sieveking G) p. 137, London, Duckworth
134. Renfrew C (1977) Alternative Models for Exchange and Spatial Distribution, in: Exchange Systems in Prehistory, (eds. Earle TK, Ericson JE) p. 71, New York, Academic Press
135. Sidrys R (1977) Mass-distance measures for the Maya Obsidian Trade, in: Exchange Systems in Prehistory, (eds. Earle TK, Ericson JE) p. 91, New York, Academic Press
136. Thorpe OW, Thorpe RS (1984) J. Archaeo. Sci. *11*: 1
137. Aspinall A, Feather SW (1972) Archaeometry *14*: 41
138. DeBruin M, Korthoven PJM, Duin RPW, Groen FCA, Bakels CC (1973) J. Radioanalyt. Chem. *15*: 181
139. Sieveking G, Craddock PT, Hughes MJ (1970) Nature *228*: 251
140. Sieveking G, Brush P, Ferguson J, Craddock PT, Hughes MJ, Cowell MR (1972) Archaeometry *14*: 151
141. Luedtke BE (1978) American Antiquity *43*: 413
142. ibid. (1979) *44*: 744
143. Heizer RF, Stross F, Hester TR, Albee A, Perlman I, Asaro F, Bowman H (1973) Science *182*: 1219
144. Bowman H, Stross FH, Asaro F, Hay RL, Heizer RF, Michel HV (1984) Archaeometry *26*: 218
145. Andrews GF (1975) Maya cities: Placemaking and Urbanization, Norman, University of Oklahoma Press
146. French JM, Sayre EV, van Zelst L (1985) Nine Medieval French Reliefs: The Search for a Proven-

    ance, in: Application of Science in Examination of Works of Art, (eds. England PA, van Helst L) p. 132, Boston, Museum of Fine Arts

147. Holmes LL, Little CT, Sayre EV (1986) J. Field Archaeology *13*: 419

148. Meyers P, van Zelst L (1977) Radiochimica Acta *24*: 197

149. Renfrew C, Springer Peacy J (1968) J. Brit. School of Archaeology at Athens *63*: 45

150. Rybach L, Nissen VU (1965) Neutron Activation of Mn and Na Traces in Marbles Worked by the Ancient Greeks, in: Radiochemical Methods of Analysis, Vol. I, p. 105, Vienna, Internat. Atomic Energy Agency

151. Lazzarini L, Moschini G, Stievano BM (1980) Archaeometry *22*: 173

152. Conforto L, Felici M, Monna D, Serva L, Taddeuci A (1975) ibid. *17*: 201

153. Craig H, Craig V (1972) Science *176*: 401

154. Manfra L, Masi U, Turi B (1975) Archaeometry *17*: 215

155. Germann K, Holzmann G, Winckler FJ (1980) ibid. *22*: 99

156. Herz N (1985) Isotopic Analysis of Marble, in: Archaeological Geology, (eds. Rapp Jr G, Gifford JA) p. 331, New Haven, Yale Univ. Press

157. Herz N, Kane SE, Hayes WB (1985) Isotopic Analysis of Sculpture from the Cyrene Demeter Sanctuary, in: Application of Science in Examination of Works of Art, (eds. England PA, van Zelst L) p. 142, Boston, Museum of Fine Arts

158. Kohl PL, Harbottle G, Sayre EV (1979) Archaeometry *21*: 131

159. Luckenbach AH, Holland CG, Allen RO (1975) ibid. *17*: 69

160. Allen RO, Pennell SE (1978) Rare Earth Element Distribution Patterns to Characterize Soapstone Artifacts, in: Archaeological Chemistry II, (ed. Carter GF) p. 230, Washington, D.C., American Chemical Society

161. Moffatt D, Butler SJ (1986) Archaeometry *28*: 101

162. Similar difficulties were experienced by P. J. Betancourt in attempting to apply the rare-earth profile method to steatite deposits found in Crete (private communication)

163. Warashina T, Kamaki Y, Higashimura T (1978) J. Archaeo. Science *5*: 283

164. Warren SE (1979) Archaeophysika *10*: 316

165. Gordus AA, Griffin JB, Wright GA (1971) Activation Analysis Identification of the Geologic Origins of Prehistoric Obsidian Artifacts, in: Science and Archaeology, (ed. Brill RH) p. 222, Cambridge, MIT Press

166. Cobean RH, Coe MD, Perry Jr EA, Turekian KK, Kharkar DP (1971) Science *174*: 666

167. Anderson DC, Tiffany JA, Nelson FW (1986) American Antiquity *51*: 837

168. Asaro F, Michel HV, Sidrys R, Stross FH (1978) ibid. *43*: 436

169. Cann JR, Dixon JE, Renfrew C (1968) Scientific American *218*: 38

170. Charlton TH, Grove DC, Hopke PK (1978) Science *201*: 807

171. Ericson JE, Kimberlin J (1977) Archaeometry *19*: 157

172. Boksenbaum MW, Tolstoy P, Harbottle G, Kimberlin J, Neivens M (1987) J. Field Archaeo. *14*: 65

173. Glascock MD, Cobean RH (1968) Trans. Am. Nucl. Soc. *53*: 198

174. Bowman HR, Asaro F, Perlman I (1973) Archaeometry *15*: 123

175. Neivens M, Harbottle G, Kimberlin J (1981) unpublished research

176. Sigleo AC (1975) Science *189*: 459

177. Ruppert H (1983) Berliner Beiträge zur Archäometrie *8*: 101

178. Ruppert H (1982) Baessler-Archiv, Neue Folge *XXX*: 69

179. Carmichael E (1970) Turquoise Mosaics from Mexico, London, The British Museum

180. Ives DJ (1984) Neutron Activation Analysis Characterization of Selected Prehistoric Chert Quarrying Areas, Ph.D. Dissertation, Columbia, Univ. of Missouri

181. Ives DJ (1986) Trans. Am. Nucl. Soc. *53*: 198

182. Sayre EV, Dodson RW (1971) Amer. Jour. Archaeology *61*: 35

183. Sayre EV, Murrenhoff A, Weick CF (1958) The Nondestructive Analysis of Ancient Potsherds Through Neutron Activation, in: Proceedings of the Boston Museum of Fine Arts Seminar, September 1958, (ed. Young WJ) p. 153

184. Abascal-M R, Harbottle G, Sayre EV (1974) Correlation Between Terracotta Figurines and Pottery from the Valley of Mexico and Source Clays by Activation Analysis, in: Archaeological Chemistry, (ed. Beck CW) p. 81, Washington, Amer. Chem. Soc.

185. Perlman I, Asaro F (1971) Pottery Analysis by Neutron Activation, in : Science and Archaeology, (ed. Brill RH) p. 182, Cambridge, MIT Press
186. The VAX version of FMTCHANGE, XFMTCHANG, and other linked taxonomic and statistical programs are available from Dr. R. L. Bishop, SARCAR Coordinator, Conservation-Analytical Laboratory, Museum Support Center, Smithsonian Institution, Washington, D.C. 20560. Phone 202-287-3715
187. Sneath PHA, Sokal RR: op. cit., pp. 305–306
188. Harbottle G (1982) Provenience Studies Using Neutron Activation Analysis: The Role of Standardization, in: Archaeological Ceramics, (eds. Olin JS, Franklin AK) p. 6, Washington, Smithsonian Institution
189. Yeh SJ, Harbottle G (1986) J. Radioanalyt. Nucl. Chem. *97*: 279
190. Ravesloot JC: Research analyses of turquoise by neutron activation, produced under contract by Harbottle G, unpublished
191. McGovern PE, Harbottle G, Wnuk C: Ware Characterization: Petrography, Chemical Sourcing, and Firing, in: The Late Bronze and Early Iron Ages of Central Transjordan, (ed. McGovern P) Chapt. 5, p. 186, Philadelphia, Univ. Pennsylvania (1986)
192. Fillieres D (1978) Contribution a l'etude de la production et l'exportation des amphores dites Marseilleses, Thesis, Doctorat 3e cycle, Paris, Université de Paris I, Pantheon-Sorbonne
193. Wolff SR, Harbottle G (1984) unpublished

Chemical Reactions Induced
and Probed by Positive Muons

Yasuo Ito

Research Center for Nuclear Science and Technology, University of Tokyo, Tokai, Ibaraki, 319-11
Japan

Table of Contents

Yasuo Ito

The application of μ^+ science, collectively called μSR, but encompassing a variety of methods including muon spin rotation, muon spin relaxation, muon spin repolarization, muon spin resonance and
level-crossing resonance, to chemistry is introduced emphasizing the special aspects of processes which
are "induced and probed" by the μ^+ itself. After giving a general introduction to the nature and
methods of muon science and a short history of muon chemistry, selected topics are given. One concerns
the usefulness of muonium as hydrogen-like probes of chemical reactions taking polymerization of
vinyl monomers and reaction with thiosulphate as examples. Probing solitons in polyacetylene induced
and probed by μ^+ is also an important example which shows the unique nature of muonium. Another
important topic is "lost polarization". Although this term is particular to muon chemistry, the chemistry underlining the phenomenon of lost polarization has an importance to both radiation and hot
atom chemistries.

1 Introduction

It was more than 50 years ago that Yukawa predicted the existence of a new particle, a meson, that should mediate nuclear forces. Before such particle was found in cosmic rays, however, another new particle was discovered. It was quite unexpected and Yukawa himself at first mistook it for his predicted particle. This new particle was initially called a μ meson, but is now known as a muon. Muons have peculiar but useful properties as is described in the following sections and, thanks to the recent development of high energy accelerators, can be obtained in copious amounts. Thus, they have become an important tool in many fields of physical chemistry and materials science.

The purpose of this article is to describe how muons can be used in chemistry. At first, a brief guide to μ^+ science is given in Sects. 2 to 4. Since there are already many excellent general review articles in muon physics and chemistry [1–7], the reader is referred to them for further details. The present review considers selected topics from a specific viewpoint: the study of reaction kinetics "induced and probed" by positive muons. Although muons are often regarded as an excellent microscopic probe which do not substantially disturb the substance to be probed, they can never be placed in it softly. They are injected with high energy and produce transient species (radiolysis products) by ionizations and excitations. Muons can react with such radiolysis products resulting in radiation chemically synthesized species. On the other hand, muons are by themselves reactive exotic particles and thus they often form a new system with the substrate. In either case, the chemical information obtained from the muons may not always be that of the undisturbed substrate but that of the system "induced" by the muons themselves. This situation is in many respects related to investigations of hot atom chemistry and radiation chemistry.

2 Properties of the Positive Muon and Muonium [3, 5–8]

All particles are divided into two groups: quarks and leptons. Leptons include the electron, muon, tauon, and the corresponding neutrinos; e^-/v_e, μ^-/v_μ, and τ^-/v_τ, and their anti-particles; $e^+/\bar{v}_e$, $\mu^+/\bar{v}_\mu$, and $\tau^+/\bar{v}_\tau$. Muons are believed to have the same properties as the electrons, except for the different mass and magnetic moment (Table 1). The muon mass (106 MeV/c^2) is 207 times larger than the electron mass (0.511 MeV/c^2), but a little smaller than the pion mass (140 MeV/c^2). Thus the muons are readily produced when charged pions decay, with a lifetime of 26 ns, by the weak interaction,

$$\pi^+ \rightarrow \mu^+ + v_\mu \tag{1}$$

In weak interactions, where neutrinos are involved, the symmetric property (i.e. the parity) is not conserved. This is related to the fact that the neutrino spin is 100% polarized anti-parallel to its momentum (helicity $= -1$) since the neutrino mass is nearly zero. (Helicity of the anti-neutrino is $+1$ however, i.e. its spin is

Yasuo Ito

Table 1. Properties of the positive muon compared with e^- and p^+

		μ^+	e^-	p^+
mass	(m_e)	206.77	1.0	1836.3
	(MeV/c^2)	105.66	0.511	938.08
charge		$+e$	$-e$	$+e$
spin		1/2	1/2	1/2
magnetic moment (μ_e)		0.004836	1.0	1.519×10^{-3}
lifetime		2.1971 μs	∞	∞

parallel to the momentum.) The pion spin is zero. Thus, in the rest frame of the pion, μ^+ must have an opposite spin to the ν_μ spin and must fly in the opposite direction. An important consequence is that the helicity of μ^+ is the same as that of ν_μ, i.e. μ^+ spin is 100% polarized anti-parallel to the direction of the momentum. μ^+ further decays into e^+ and neutrinos with the lifetime of 2.2 μs.

$$\mu^+ \rightarrow e^+ + \nu_e + \tilde{\nu}_\mu \,. \tag{2}$$

This decay is induced by the weak interaction and, as the result of parity non-conservation again, e^+ is emitted preferentially in the μ^+ spin direction. The probability R of e^+ emission at an angle θ with respect to the spin direction is proportional to:

$$R(\theta) = 1 + \tilde{A}_0 \cos \theta \,, \tag{3}$$

where $\tilde{A}_0$ is the spin asymmetry which, averaged over all possible μ^+ energies, becomes about 1/3. It is due to these asymmetric properties that muons have found their unique applications.

Although muons are thought to have the same properties as electrons, as is indeed the case for negative muons, μ^-, which behave as "heavy electrons",[9] the chemical properties of positive muons should rather be regarded as "light protons". The positive muon can pick up an electron from a substance and form a neutral particle called "muonium" ($\mu^+ e^-$, chemical symbol: Mu). The atomic parameters of Mu are very close to those of H atom except that the mass is 9 times smaller and it is not stable because of the intrinsic decay nature of μ^+. Mu can be regarded as a radioactive isotope of H atom (Table 2).

Table 2. Properties of muonium compared with H atom

		Mu	H
mass	(m_e)	207.8	1837.2
reduced mass	(m_e)	0.9952	0.9994
Bohr radius	(nm)	0.05315	0.05292
ionization potential	(eV)	13.54	13.60
hyperfine frequency	(MHz)	4463	1420

Mu is a typical two $^1/_2$-spin system and the spin Hamiltonian is given by,

$$H = a\vec{I} \cdot \vec{S} + g_e\mu_e\vec{S} \cdot \vec{B} - g_\mu\mu_\mu\vec{I} \cdot \vec{B} \tag{4}$$

where a is the hyperfine coupling constant between μ^+ and e^-, g_e, g_μ and μ_e, μ_μ are the g-factors and magnetons of e^- and μ^+, respectively, S and I are the electron and muon spin momenta, and B is the magnetic field intensity. It is convenient to rewrite the above equation as,

$$H = \frac{\hbar}{2}\omega_e\sigma_e + \frac{\hbar}{2}\omega_\mu\sigma_\mu + \frac{\hbar}{4}\omega_0(\sigma_\mu\sigma_e) \tag{5}$$

where ω_0 is the hyperfine (hf) frequency, $\omega_e = 2g_e\mu_eB/\hbar$ and $\omega_\mu = -2g_\mu\mu_\mu B/\hbar$ are the Lamor precession frequencies of e^- and μ^+, respectively, σ_e and σ_μ are the Pauli spin operators, and $\hbar$ is Planck's constant. The eigenstates are described using the spin quantum numbers of μ^+ (m_μ) and e^- (m_e) as,

$$|1\rangle = |\alpha_\mu\alpha_e\rangle$$

$$|2\rangle = s|\alpha_\mu\beta_e\rangle + c|\beta_\mu\alpha_e\rangle$$

$$|3\rangle = |\beta_\mu\beta_e\rangle$$

$$|4\rangle = c|\alpha_\mu\beta_e\rangle - s|\beta_\mu\alpha_e\rangle \tag{6}$$

where

$$s = \frac{1}{\sqrt{2}}\left(1 - \frac{x}{\sqrt{1 + x^2}}\right)^{1/2}$$

$$c = \frac{1}{\sqrt{2}}\left(1 + \frac{x}{\sqrt{1 + x^2}}\right)^{1/2}$$

$$x = 2\omega_+/\omega_- = B/1585G, \quad \omega_\pm = \frac{\omega_e \pm \omega_\mu}{2}$$

x is the magnetic field intensity divided by the internal magnetic field of Mu. The energy levels of the eigenstates are:

$$\omega_1 = \omega_- + \frac{\omega_0}{4}$$

$$\omega_2 = -\frac{\omega_0}{4} + \left(\omega_+^2 + \frac{\omega_0^2}{4}\right)^{1/2}$$

$$\omega_3 = -\omega_- + \frac{\omega_0}{4}$$

$$\omega_4 = -\frac{\omega_0}{4} + \left(\omega_+^2 + \frac{\omega_0^2}{4}\right)^{1/2} \tag{7}$$

The energy levels are shown in the form of the Breit-Rabi diagram in Fig. 1. At low magnetic fields, the total angular momentum F and M_F are good quantum numbers. At high fields, however, the individual spin quantum numbers m_μ and m_e describe the system.

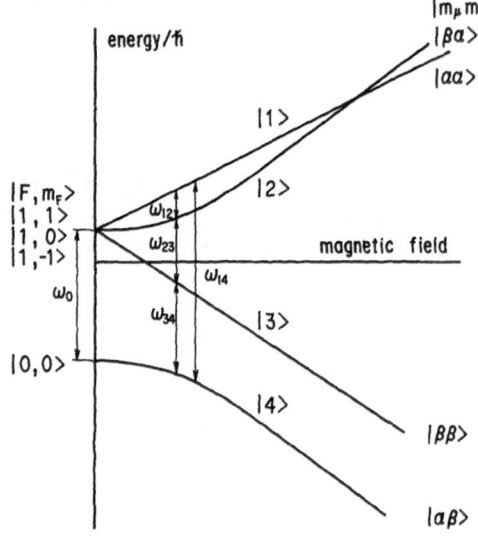

Fig. 1. The energy levels of the muonium spin states as a function of the magnetic field. At low field intensity the total angular momentum F and the magnetic quantum number M_F are good quantum numbers, while at high fields the individual spin quantum number of μ^+ and e^- describe the system. The *arrows* indicate the allowed transitions which correspond to muonium spin rotation and muonium spin resonance

As has been explained, the μ^+ spin is fully polarized. If this spin state is specified by $|\alpha_\mu\rangle$, Mu is formed either as $|\alpha_\mu\alpha_e\rangle$ or $|\alpha_\mu\beta_e\rangle$ with equal probability. $|\alpha_\mu\alpha_e\rangle$ is one of the eigenstates of triplet Mu (Eq. 6) and is customarily called triplet-Mu (^{T}Mu). Although referred to as singlet-Mu (^{S}Mu), $|\alpha_\mu\beta_e\rangle$ is not an eigenstate at low field; it is a mixture of two m = 0 states ($|1,0\rangle$ and $|0,0\rangle$), and the spin state oscillates between them with a hyperfine frequency of 4463 MHz (Table 2). At high field, however, $|\alpha_\mu\beta_e\rangle|$ reaches the eigenstate $|4\rangle$.

Positive muons injected into matter are at first decelerated by Bethe-Bloch type energy loss processes. When their velocity becomes close to that of the valence electrons, it becomes possible for μ^+ to pick up an electron and form energetic Mu, which, on subsequent collision, will be dissociated.

$$\mu^+ + M \rightarrow Mu^* + M^+ \tag{8}$$

$$Mu^* + M \rightarrow \mu^+ + e^- + M .$$

This charge-exchange cycle is repeated until an energy close to the ionization potential of the substance is reached, whereupon the cycle is terminated. At this point some positive muons may already be incorporated into diamagnetic species by hot atom reactions, while the rest remain as epithermal free μ^+ or epithermal Mu (Fig. 2). It is important to know what fraction of μ^+ leaves the charge-exchange cycles as free μ^+ or as Mu, because this fraction represents the chemical consequences of μ^+ in the succeeding reactions. It is generally accepted that in substances having an ionization potential larger than that of Mu ($= 13.6$ eV), free μ^+ will be the main product of the charge-exchange cycles. In substances having a smaller ionization potential, more μ^+ will leave the cycles as Mu. A theoretical estimation shows that in hydrocarbons 80% of μ^+ comes out as Mu and 20% as free μ^+ [10].

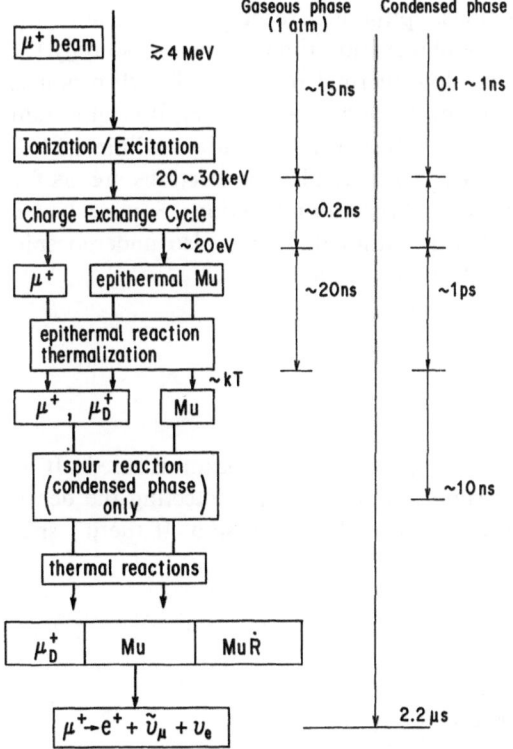

Fig. 2. Chemical processes of μ^+ injected into molecular substances with an energy larger than about 4 MeV, the lowest energy of the μ^+ beam usually available at muon facilities. μ_D^+ and MuṘ are diamagnetic muons and Mu-substituted radicals, respectively. The right side shows approximate time scales in gaseous and condensed phases.

Reactions of epithermal μ^+ and epithermal Mu also induce changes in the chemical state of μ^+ and, after thermalization, μ^+ will emerge incorporated either in diamagnetic states (with a fraction h), muonium (fraction m) or Mu-substituted free radicals (fraction r). They may further change their chemical states either by thermal reactions or, in the condensed phases, by intra-spur reactions.

"Diamagnetic muons" (denoted by μ_D^+) are the compounds that include μ^+ but not unpaired electrons. They have the chemical form in which Mu has replaced to H atom in diamagnetic molecules. MuH, MuOH, and MuCl are the examples. The magnetic property of the diamagnetic muon is almost the same as the bare μ^+ since there is no strong internal magnetic coupling in the molecule. Solvation of μ^+, abstraction and exchange reaction of Mu, and Mu addition to free radicals lead to diamagnetic muons. "Mu-substituted radicals" are the μ^+ compounds that include an unpaired electron magnetically coupled to the μ^+ spin. Mu addition to unsaturated bonds leads to Mu-substituted free radicals formation [11–13]. But the pathway to the formation of the Mu-substituted free radicals is not limited to simple Mu addition [14].

It is important to note the time scale of the muonic processes described above. The charge-exchange cycles come to an end within several ns (in gases) or several tens of ps (in the condensed phases) after μ^+ injection. The epithermal reactions in the condensed phases may take place for only several ps, while the spur reactions extend to several ns. The time window of observation is between several ns and several µs, being limited by the time resolution of the e^+ counter (≤ 1 ns) and by the μ^+ lifetime. Thus hot atom reactions, epithermal reactions and most of the intra-spur reactions are finished before the observation time is reached, so only the products of these reactions are observed. Slow thermal reactions of such products can be measured from the relaxation of the corresponding asymmetry. It is important that ^{S}Mu is not observed at this stage because it has been depolarized within 0.22 ns ($=1/4463$ MHz) by the hf relaxation (cf. Sect. 3). Possible Mu reactions are, as for H atom, abstraction, replacement, addition, and oxidation reactions. Spin exchange is also an important reaction of Mu [15–17]. Although H can also undergo spin exchange reaction, it cannot be studied with such ease as Mu.

3 Techniques of µSR

In all experiments of µSR, a longitudinally polarized μ^+ beam is used. It is stopped in the sample and the μ^+ spin state is monitored by detecting the decay e^+ that is preferentially emitted in the direction of the muon spin. If the μ^+ spin

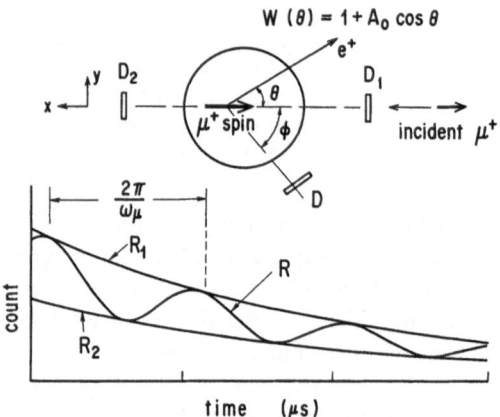

Fig. 3. The principle of measurements of μ^+ spin relaxation and rotation. The μ^+ beam enters the sample from the right side. The *heavy arrow* indicates the μ^+ spin direction which is opposite to the direction of the momentum. D_1, D_2 and D are detectors of the decay positrons. The curves R_1 and R_2 are the positron counting histograms obtained by the counters D_1 and D_2 when μ^+ spin does not evolve. When a magnetic field is applied perpendicular to the μ^+ spin (in the direction of the z axis), the spin evolves in the xy plane. This evolution, observed by the positron counter D, is the cosine function inscribed in the decay curves R_1 and R_2

polarization is not affected, one may observe e^+ emission distributed according to Eq. 3, superimposed on the decay curve of 2.2 μs.

$$R(t, \theta) \approx [1 + A_0 \cos \theta] \exp\left(-\frac{t}{2.2\mu s}\right) \tag{9}$$

For e^+ counters D_1 and D_2 placed at the upstream and downstream positions of the muon beam (Fig. 3), the e^+ counting histogram is proportional to $1 + A_0$ for D_1 and $1 - A_0$ for D_2.

If, on the other hand, the μ^+ spin is made to rotate either by an applied external field or by an internal magnetic field, the direction of the decay e^+ emission evolves. It may be compared to a light house whose illumination direction is rotating. The probability of decay μ^+ detection is then,

$$R(t) \approx [1 + A_0 \cos(\omega t + \phi)] \exp\left(-\frac{t}{2.2\mu s}\right) \tag{10}$$

where ω is the angular frequency which is determined by the gyromagnetic ratio and the magnetic field intensity. A constant angle ϕ describes the angle position of the e^+ counter D. The e^+ counting histogram (R) measured by the counter D (Fig. 3) will then be a cosine function inscribed between the maximum (R_1) and the minimum (R_2) counting curves. For a general case in which there are several different μ^+ containing species, each evolving and undergoing spin relaxation at different rates, Eq. 10 can be extended as,

$$R(t) \approx \left[1 + \sum_{i=1}^{N} A_{0i} G_i(t) \cos(\omega_i t + \phi_i)\right] \exp\left(-\frac{t}{2.2\mu s}\right) \tag{11}$$

where A_{0i} and $G_i(t)$ are the initial asymmetry and the relaxation function of the i-th component, respectively. The form of the relaxation function depends on the mechanism of relaxation, but for many cases of chemical interest it is usually given by an exponential function (Eq. 17). The term in the square brackets of Eqs. 9—11 is called "assymmetry spectrum" and is derived from the observed e^+ counting histogram simply by compensating for the μ^+ lifetime of 2.2 μs. It is customary to present the time histogram in the form of the asymmetry spectrum.

The term "μSR" was originally used for "muon spin rotation", but muon spin science has now developed into a variety of techniques and is not limited to the "rotation" technique. It is fortunate, however, that most of the other techniques (muon spin relaxation, muon spin repolarization, and muon spin resonance) have the same abbreviated form μSR. Thus μSR is sometimes used to describe all these techniques collectively. When it is necessary to specify these different techniques, "μS Rotation", "μS Resonance", etc. will be used.

3.1 μS Rotation

In μS Rotation, an external magnetic field B is applied perpendicular to the direction of the incident μ^+ spin. If the x axis is taken parallel to the incident μ^+

spin direction, and z axis parallel to the magnetic field direction, the μ^+ spin in any chemical state evolves in the xy plane. The chemical state of μ^+ can be identified from the difference in the rotation frequency. The μ^+ spin in diamagnetic muon, μ_D^+, evolves with a Lamor frequency,

$$\omega_\mu = \gamma_\mu B, \qquad \gamma_\mu/2\pi = 13.5544 \text{ kHz/G} \tag{12}$$

where γ_μ is the gyromagnetic ratio of μ^+. The ^{T}Mu evolution corresponds to the transitions between the states $|1\rangle \leftrightarrow |2\rangle$ and $|2\rangle \leftrightarrow |3\rangle$ (Fig. 1). These two transition frequencies are almost the same (degenerate) at very low magnetic field, and the corresponding frequency is much faster than that of μ_D^+ evolution.

$$\omega_M = \gamma_M B, \gamma_M/2\pi = -1.394 \text{ MHz/G} \tag{13}$$

where γ_M is the gyromagnetic ratio of Mu and the negative sign indicates that the μ^+ spin in ^{T}Mu evolves in the direction opposite to that of μ^+ evolution. At slightly larger fields, the two frequencies become different (generate), and this leads to an apparent relaxation or a beating of the muonium spin rotation asymmetry spectrum [5, 18, 19]. The evolution of μ^+ spin in ^{S}Mu corresponds to the transitions $|1\rangle \leftrightarrow |4\rangle$ and $|3\rangle \leftrightarrow |4\rangle$ and is much faster (4463 MHz) than that of ^{T}Mu.

Figure 4-a illustrates how μ^+ spin in ^{T}Mu and ^{S}Mu evolves in a transverse magnetic field. The frequency of ^{S}Mu evolution is at the limit of the time resolution and is not observed. The ^{S}Mu polarization looks as if depolarized, and this is called "hf relaxation", i.e. the relaxation that occurs at the rate of the hf frequency. Figure 4-b is a typical asymmetry spectrum of μS Rotation measured at 18G, where the evolutions of ^{T}Mu and diamagnetic muon are observed superimposed. (Note the difference of the time scale from Fig. 4a). The fast damping of the ^{T}Mu asymmetry is apparent, and is caused by the two slightly different frequencies of ^{T}Mu precessions. Thus it is customary to measure the ^{T}Mu evolution in a much lower field (≈ 3G) in which such a two-frequency splitting is not significant. The precession of the diamagnetic muons is usually measured around 100G to see more μ_D^+ rotations.

At high magnetic fields, the two transitions $|1\rangle \leftrightarrow |2\rangle$ and $|3\rangle \leftrightarrow |4\rangle$ (Fig. 1) are in the range of observable frequencies, and the corresponding frequencies are,

$$\omega = \left| \omega_\mu \pm \frac{\omega_0}{2} \right| \tag{14}$$

The evolution of μ^+ spin in Mu-substituted free radicals is complicated owing to the splitting of energy levels caused by super-hyperfine coupling of the nuclear spins. A detailed theoretical treatment of Mu-substituted free radicals has been given.[11] At sufficiently high fields, however, the super-hyperfine interactions are decoupled (Paschen-Back effect) and the μ^+ spin evolves according to the hf coupling with the unpaired electron spin. In such a case the Mu-substituted radical is regarded as a "pseudo-Mu" atom with a larger $\mu^+ - e^-$ distance (hence smaller hf constant) than in Mu. Eq. 14 applies to the precession of Mu-substituted radicals in the Paschen-Back regime too, and different Mu-substituted free radicals can be identified by the differences in the hf constant ω_0. The technique of observing Mu-substituted radicals in this way is called "high-field μSR". The raw histogram

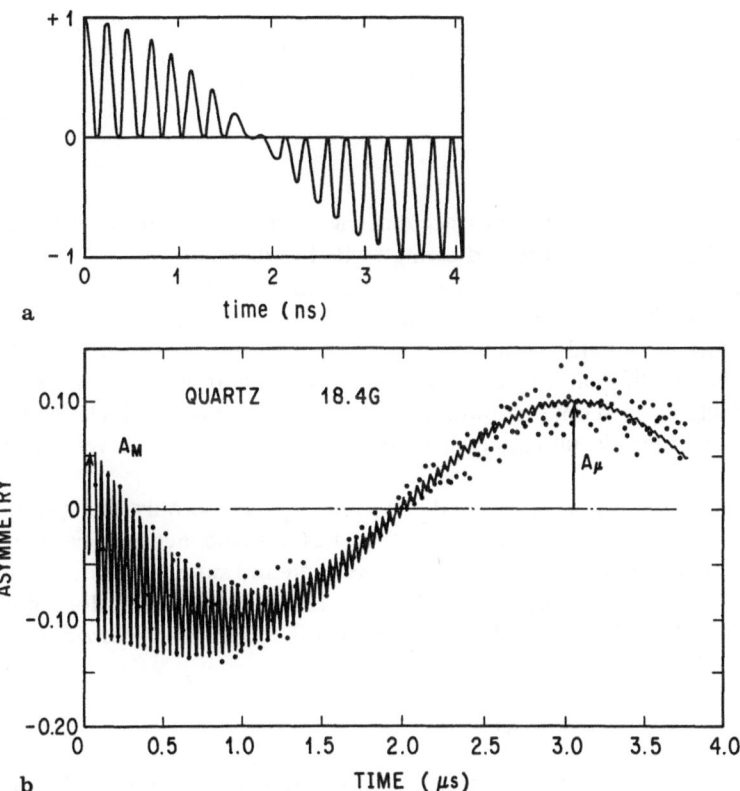

a

b

Fig. 4.a: Theoretical evolution of Mu in a transverse magnetic field. The fast oscillation is the hyperfine oscillation of ^{S}Mu and the slow oscillation, showing only half of the period, is the rotation of ^{T}Mu.
b: An example μS Rotation spectrum of quartz measured at 18.4 G transverse magnetic field. The fast oscillation is the rotation of ^{T}Mu and the slow rotation is that of μ_D^+. The amplitude of each oscillation corresponds to the asymmetry of the corresponding species

(corresponding to Eq. 11) is a superposition of several high frequency rotations and, by Fourier transformation power spectrum, the frequencies and the corresponding asymmetries are derived.

In principle every muon state can be identified by the frequency of μS Rotation, and the intensity of each state is determined from the corresponding asymmetry (Eq. 11). The fraction of each component is directly given by the polarization. The polarizations for the diamagnetic muon (P_D), muonium (P_M), and Mu-substituted

103

radicals (P_{Ri}), are defined by dividing the corresponding asymmetries by the total asymmetry, $\tilde{A}$, of the incident μ^+ beam.

$$P_D = \frac{A_{0D}}{\tilde{A}},$$

$$P_M = 2 \times \frac{A_{0M}}{\tilde{A}},\qquad\qquad\qquad (15)$$

$$P_{Ri} = \frac{A_{0Ri}}{\tilde{A}}.$$

The factor 2 for P_M arises to take into account ^SMu which is formed with the same amount as ^TMu but is not observed owing to its hf relaxation.

It is unfortunate that μSR does not have the resolution to distinguish different diamagnetic muons such as MuH, MuOH and MuCl in the way NMR does by the chemical shift. The field inhomogeneity associated with a not very small μ^+ beam size makes it difficult to determine the precession frequency precisely. Even if such obstacle is overcome in future, the intrinsic linewidth, which is due to the definite μ^+ lifetime, is ultimately left and the resolution can never reach the level of NMR.

It must be noted that the polarizations P_D, P_M and P_R include only the species which have coherent μ^+ spin. Any μ^+ spin that has become out of phase or depolarized does not contribute to the coherent precession and simply merges in the background. Thus quite often, the sum of the observed polarization does not amount to unity, and the polarization that has been lost is called "lost polarization" or "missing fraction", P_L.

$$P_L = 1 - (P_D + P_M + P_R).\qquad\qquad\qquad (16)$$

In low density gases the lost polarization may come from hf relaxation of ^SMu during the charge-exchange cycle, if the time between each collision is longer than the hf relaxation time 0.22 ns [5, 20]. But in condensed phases, the time spent in the charge-exchange cycles and in the epithermal stage is short, and the spin exchange of ^TMu with free radicals (hydrated electrons [21], H atoms, etc.) and slow reactions of Mu are considered to be the important sources of the lost polarization. The mechanism of the lost polarization is not well understood, and remains as an interesting subject in conjunction with radiation chemistry.

Any μ^+ spin relaxation that occurs within the time window of μSR (several ns $\sim$ several tens of μs) can be measured precisely (Eq. 11). When ^TMu undergoes a chemical reaction or spin exchange reaction with an additive X, the Mu relaxation function is described by a quasi first-order decay as

$$G_M(t) = \exp\left[-(\lambda_0 + k_M[X])\, t\right]\qquad\qquad\qquad (17)$$

where λ_0 is the intrinsic decay rate of ^TMu. This method of determining the rate constant k_M is known as "Mu decay kinetics". Measurements of the rate constants

k_M both in gases [22] and liquids [23–27] are the subjects of common interest for chemists, because comparisons can be made with those of H atoms, k_H. This provides a basis for the study of kinetic isotope and tunneling effects [28–30].

The diamagnetic polarization P_D has at least two origins and it is normally described as

$$P_D = h + mP_{res} \tag{18}$$

where h, temperature independent and little affected by a small amount of additives, is the fraction of diamagnetic muons produced in hot or epithermal reactions or fast spur processes. P_{res}, the residual polarization, is the fraction of Mu (initially formed with a fraction m) that is incorporated into diamagnetic compounds by reactions which are slower than hot and spur reactions, but fast enough so that μ^+ does not lose spin coherency. Suppose a case where ^TMu and ^SMu can become diamagnetic muons by reaction with substance RX;

$$^TMu/^SMu + RX \rightarrow MuX \ (diamagnetic) + R \ . \tag{19}$$

Whether or not the μ^+ spin in diamagnetic muon MuX retains the spin coherency is determined as the result of competition between the spin evolution frequency and the reaction rate $\lambda = k[RX]$: if Mu reacts faster before its spin becomes out of phase by evolution, the spin coherency will be retained. Theoretical treatment of this problem is straightforward [5, 18], but only the result is quoted here. The field dependence of Pres is illustrated in Fig. 5-a for various reaction rates λ given in the unit of the hf relaxation frequency, ω_0. Apparently P_{res} can be divided into two parts. Below about $\lambda = 0.3 \, \omega_0$, ^TMu is the main source of P_{res}, which has a strong field dependence and decreases steeply at higher fields. This is because ^TMu spin evolution becomes faster at higher fields and more and more μ^+ spins become out of phase in reaction 19. Above $\lambda = 0.5 \, \omega_0$, the reaction is fast enough and most of the ^TMu is brought into coherent diamagnetic muon states by the reaction. At such a moderately high reaction rate, ^SMu is still depolarized by the hf relaxation rate. But even the hf relaxation is intercepted when the reaction rate becomes much larger. The dependence of P_{res} and its phase on λ gives an experimental basis for the study of reaction 19. A similar treatment also applies when Mu-substituted radicals are the cause of the loss of polarization [31].

Figure 5-b illustrates how the residual polarization is treated experimentally for a case where the diamagnetic muon polarization was measured as function of concentration of additives ($RX = I_2$) in methanol [5]. The rise of the polarization is due to reaction 19, where MuX = MuI. The first plateau near $[I_2] = 0.07 \, M$ implies that ^TMu has reacted completely. Much higher I_2 concentrations are required to bring ^SMu into coherent diamagnetic muon. Since ^TMu and ^SMu evolution frequencies ω_M and ω_0 are known, it is possible to determine the reaction rate λ. Here the two evolution frequencies ω_M and ω_0 are acting as two different clocks to measure the reaction rate. This method of determining the rate constant is complementary to that of Mu decay kinetics. A comparison of these two different methods is given in Sect. 5-2.

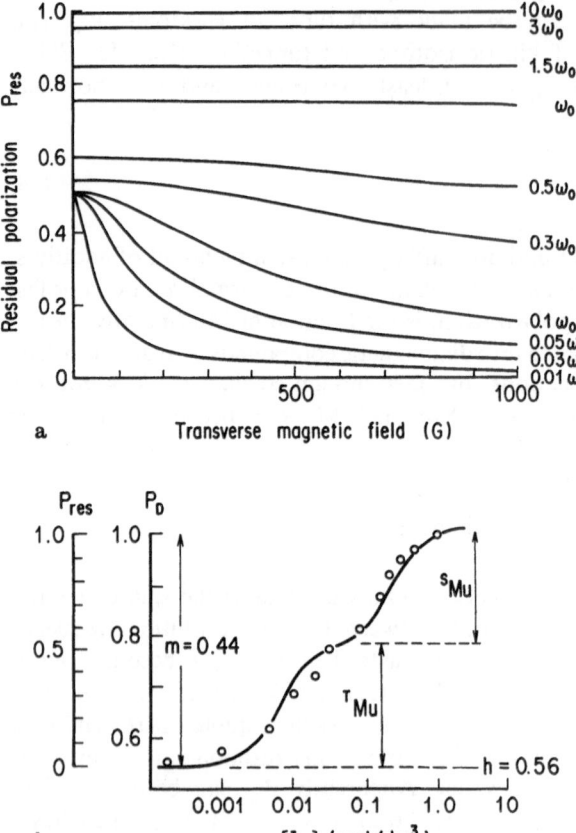

Fig. 5a, b. Residual polarization a: Theoretical residual polarization for muonium undergoing chemical reaction (19) of the text by which Mu is incorporated into diamagnetic muon. The reaction rate, λ, is given in the unit of the hyperfine frequency ω_0. ^{T}Mu becomes easily out-of-phase at low magnetic fields, and the reaction rate must be larger than c.a. $0.3\omega_0$ to incoorporate it into residual polarization. A much larger reaction rate is required to bring ^{S}Mu into coherent polarization because it depolarizes much faster due to the hyperfine frequency b: An example residual polarization experiment (see text)

3.2 μS Relaxation

In μS Relaxation, the static magnetic field is applied parallel to the initial μ^+ spin direction (x axis of Fig. 3). Where there are no magnetic interactions, the hf relaxation of ^{S}Mu and of "pseudo-^{S}Mu" will be the only source of μ^+ spin relaxation, and μ^+ spin in any other chemical states will not evolve. Where there are electric or nuclear spins that interact with μ^+ spin, the latter will start to evolve at a rate depending on the magnetic field intensity formed by the electric or nuclear spins. The experiments are performed using positron counters placed at the upstream and/or downstream positions of the μ^+ beam (D_1 and D_2 in Fig. 3). The count rates of the counters, $R_1(t)$ and $R_2(t)$, are described using Eq. 3 as,

$$R_1(t) = 1 + AG(t), \qquad R_2(t) = 1 - AG(t) \tag{20}$$

where G(t) is the relaxation function which contain information about the kind and degree of magnetic interactions.

A particular usefulness of μS Relaxation to chemistry is that it gives a means of distinguishing between two different spin relaxation types: spin depolarization and dephasing. By μS Rotation method it is not possible to distinguish between the relaxation due to depolarization, as in the spin exchange reaction

$$^{T}Mu + S \rightarrow {}^{S}Mu + S \tag{21}$$

and the relaxation due to dephasing, as in the chemical reaction

$$^{T}Mu + RX \rightarrow MuX + R . \tag{22}$$

But, using μS Relaxation it is possible to distinguish these reactions. Since μ^{+} spin in ^{T}Mu and MuX does not evolve in a longitudinal field, the asymmetry in the direction of the magnetic field does not change in reaction 22, while it does change by hf relaxation of ^{S}Mu in reaction 21.

3.3 μS Repolarization

The experimental arrangement of μS Repolarization is the same as that of μ S Relaxation, and the longitudinal polarization $P^{//} = A^{//}/\tilde{A}$ is measured as a function of the longitudinal magnetic field intensity. The superscript $//$ indicates that the polarization is defined in the longitudinal direction. At extremely high fields all the magnetic interactions, even the hyperfine interactions of ^{S}Mu, are decoupled and the μ^{+} spin polarization should recover to unity. At intermediate magnetic fields, the degree of recovery of the polarization depends on the rate of the magnetic interactions. When hf relaxation of ^{S}Mu is the only source of the spin relaxation, for example, the recovery of the polarization follows the expression,

$$P^{//}_{Mu} = \frac{1}{2} + \frac{x^2}{2(1 + x^2)} . \tag{23}$$

The first term corresponds to ^{T}Mu polarization and the second to ^{S}Mu polarization. When Mu is involved in a reacting system where there are both spin exchange (with the rate v) and chemical reactions (with the rate $1/\tau$), the repolarization effect of the longitudinal magnetic field is expressed as [32]

$$P^{//}_{Mu} = 1.0 - \frac{(\omega_0 \tau)^2 (1 + 2 v\tau)}{2[(\omega_0 \tau)^2 (1 + v\tau + x^2) + (1 + 2 v\tau)^2]} . \tag{24}$$

Figure 6 illustrates the principle of the μS Repolarization, in which it is shown how the longitudinal polarization depends on the magnetic field. When there are only ^{T}Mu and ^{S}Mu, the polarization increases from $^{1}/_{2}$ for ^{T}Mu to unity at high fields (solid curve 1). For a system having a smaller hf coupling constant, as in pseudo-Mu, the polarization reaches unity much faster (solid curve 2). From the shape of the repolarization curve, therefore, inference can be made as to what species were responsible for the loss of polarization. The broken curves in Fig. 6 drawn using Eq. 24 show that the polarization at zero field provides further information about the chemical interaction by which ^{S}Mu is incorporated into diamagnetic muon without spin relaxation.

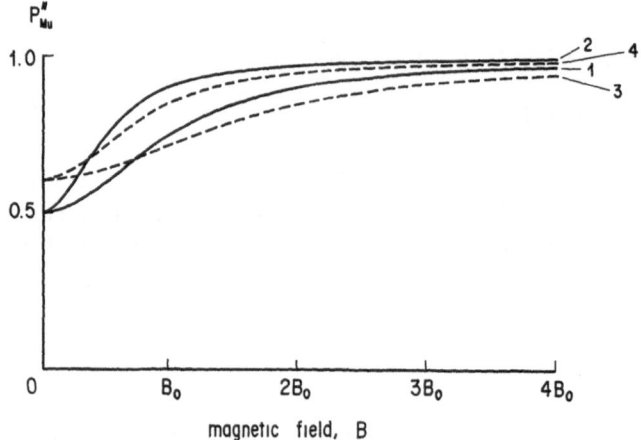

Fig. 6. μS Repolarization. Longitudinal polarization of Mu as a function of the magnetic field (unit: the internal magnetic field intensity, $B_0 = 1585$ G). *Curve 1* corresponds to Eq. 23 for Mu, and *curve 2* is that for a pseudo-Mu with a hyperfine constant smaller than that of Mu. *Broken curves 3* and *4* correspond to Eq. 24 calculated for Mu and pseudo-Mu, respectively, and they indicate that the longitudinal polarization $P_{Mu}^{//}$ is enhanced at zero field due to chemical reactions. The shape of the repolarization curve thus contains information on the species and reactions that cause the loss of μ^+ spin polarization

The overall polarization in the longitudinal field is then described as,

$$P_{total}^{//} = h + (1 - h)\, P_{Mu}^{//}\,. \tag{25}$$

The μS Repolarization method has been successfully used to make a guess of the species that caused the loss of spin polarization in various liquids [33].

3.4 μS Resonance [34]

As in NMR, muon spin resonance occurs under a longitudinal magnetic field and RF. When the RF frequency becomes equal to $\gamma_\mu H_0$ (H_0 is the applied magnetic field), the μ^+ spin starts to evolve, and the time dependence of the on-resonance component follows the expression,

$$R_u(t) = [1 + A_{res} G(t)\cos(\gamma_\mu H_1 t)]\exp\left(-\frac{t}{2.2\mu s}\right) \tag{26}$$

where A_{res} is the asymmetry of the on-resonant component, H_1 is the intensity of the RF-induced magnetic field perpendicular to the incident μ^+ spin direction.

The resonance can be monitored directly by measuring the asymmetry spectrum by one of the positron counters, and the evolution of μ^+ spin on resonance is directly observed as shown in Fig. 7a. By fitting the spectrum to Eq. 26, the asymmetry and the relaxation of the on-resonant species are obtained. By integrating the counts of the asymmetry spectrum over time, signal intensities $1 - A_{res}$ and $1 + A_{res}$ are obtained for the positron counters D_1 and D_2 (Fig. 3), respectively, and their plot as a function of the field intensity gives the resonance line profile (Fig. 7b). μS Resonance has an extreme sensitivity: only about 10^{7-8} μ^+ suffice to get one resonance lineshape spectrum, which is to be compared to 10^{13-14} spins necessary for an NMR spectrum.

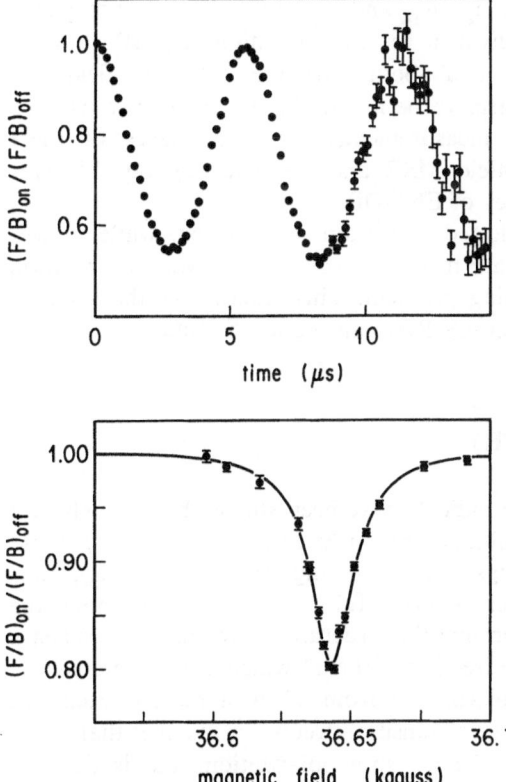

Fig. 7. Example of the μS Resonance of CCl_4. Since the diamagnetic muon polarization of CCl_4 is 1.0, it is frequently used as a reference material to calibrate the μ^+ asymmetry. The upper spectrum is the evolution time spectrum of the diamagnetic muon at just-resonance (at the magnetic field corresponding to the resonance peak shown below). This corresponds to the term in the square bracket of Eq. 26.

The lower spectrum is the resonance lineshape obtained by integration of the evolution time histogram (upper spectrum) at each magnetic field. The ordinate is the ratio of the counts of the upstream (F) and downstream (B) positron counters on resonance divided by that off resonance, and is proportional to $1 - A_{res}/1 + A_{res}$

An important difference between the Rotation and the Resonance methods is that the latter does not involve problems of dephasing and depolarization (cf. earlier discussion on the differences between the two methods). In μS Rotation, all μ^+ in any chemical states must evolve. Thus, when some secondary μ^+ state is formed by slow processes, the μ^+ spin may become out of phase with the initial spin direction. Such μ^+ spin does not give a coherent Rotation signal and simply merges into the background. In the Resonance method, on the other hand, the magnetic field is applied in the direction of the μ^+ spin and μ^+ evolution does not occur except for the on-resonant component. Furthermore, the Resonance experiments are performed using a sufficiently high longitudinal field so that all depolarization is repolarized. The problem of loss of polarization can thus be avoided. When, by slow processes, a μ^+ is converted to the on-resonant species, it can be detected directly in the asymmetry spectrum. This technique has been successfully used to study the Mu and Mu-substituted radical reactions in liquid hydrocarbons [33] and in ionic crystals [35].

3.5 Muonium Spin Resonance

Muonium spin resonance resembles ESR. As in muonium spin rotation, there are four allowed Mu spin transitions: ω_{12}, ω_{23}, ω_{14} and ω_{34} (Fig. 1). At low magnetic fields, the first two resonances are equivalent to ^{T}Mu evolution of μ SR, and the latter two correspond to ^{S}Mu evolution of about 4463 MHz. At high fields the frequencies ω_{23} and ω_{14} become too large to be measured and the two transitions ω_{12} and ω_{34} are then suitable for measurements. The frequencies of these transitions are the same as the high-field MSR and are given by Eq. 14. This equation is the same as for the frequency of ENDOR.

In Mu spin resonance only the selected transition is activated, while in Mu spin rotation all the other transitions are made to evolve. Thus Mu spin rotation is always accompanied by the dephasing problem. This situation is the same as that described for the difference between μS Resonance and μS Rotation.

3.6 Level-Crossing Resonance (LCR)

A large number of Mu-substituted free radicals have been studied by the high-field μSR, but this method only gives the values of muon hf constants and is incapable of determining other nuclear couplings. Level-crossing resonance (LCR), first suggested by Abragam [36], provides a powerful means of determining the nuclear hf constants and hence to determine the structure of free radicals. When a magnetic field is applied along the spin direction of the μ^+ which is incorporated in a free radical, a resonant transfer of polarization from μ^+ to a nucleus occurs at particular magnetic fields where the μ^+ spin transition frequency matches that of the nucleus. This results in a decrease in the μ^+ spin polarization, and is detected experimentally as a decrease in the μ^+ asymmetry along the applied field direction. For a Mu-substituted radical in solution where the μ^+ hf constant is much larger than that of the nuclei, equivalent nuclei give overlapping resonance at the central position expressed by,

$$B_R = \frac{1}{2}[(a_\mu - a_n)/(\gamma_\mu - \gamma_n) - (a_\mu + a_n)/\gamma_e] \qquad (27)$$

where a_μ, a_n and γ_μ, γ_n are the hf and gyromagnetic ratios for μ^+ and nucleus, respectively.

The advantage of LCR as compared to high-field μ SR is not only in its capability of determining nuclear hf couplings. Since the magnetic field is applied in the direction of the initial μ^+ spin, there is no problem with dephasing caused by slow reactions, for the same reason as for the Resonance method.

4 Short History of Muon and Muonium Chemistry

The value of μSR as physical and chemical probes became apparent soon after 1957 when the violation of parity conservation was discovered for π^+ and μ^+ [37]. The first observations of ^TMu were carried out in 1960 in gases,[1] and the studies of the gas phase Mu reactions subsequently followed [38]. Extensive data on diamagnetic muon polarization in condensed phases were accumulated, and all the data were analyzed assuming the existence of Mu which, by reaction with substrates, was supposed to be incorporated into diamagnetic muons (by addition or replacement reactions) or into radicals (by addition to unsaturated bonds). Direct observation of ^TMu in liquids had to wait until the end of the 1970s when it was observed in water [39], alcohols [40], and saturated hydrocarbons [41]. One of the problems of common interest to chemists was how stable Mu is in water, since the "intrinsic" ^TMu lifetimes in water varied according to the researchers. The actual "intrinsic" ^TMu lifetime was carefully measured after eliminating all possible disturbances, impurities and inhomogeneity of the magnetic field, and using a pulsed muon beam that enabled measurements of MSR (Muonium Spin Rotation) for a longer time window than before. As a result ^TMu was found to be stable on the time scale of several tens of μs [42]. Owing to the long chemical lifetime of Mu in water, it is possible to measure ^TMu reaction rates with various additives (Eq. 17), with the technique of Mu decay kinetics. The Mu reaction rates k_M thus measured by the Mu decay kinetics can be compared with those of the H atom, k_H. For aqueous solutions, the ratio k_M/k_H ranges from 10^{-2} to 10^2. Discussion has been made with regard to activation process [43], diffusion kinetics [6], and spin exchange reactions [6, 15]. MSR study in micellar systems has also been made [44].

Mu-substituted free radicals had not been detected by μSR at about 100 G which had been the magnetic field intensity commonly used in the muon facilities, and the first observation was made at high transverse magnetic fields (about 3000 G) in 1978 [11, 45] and this method has sometimes been called "high-field μSR". A great number of radicals have been measured since then, e.g. olefins and dienes [46], methyl-, F- and other substituted benzenes [13, 47], and triple bonds [48]. All the observed radicals are derived by Mu addition to unsaturated molecules, and thus μ^+ is automatically located at the β-position, i.e. two bond, away from the unpaired electron or delocalized system. Although Mu-substituted radicals are typical entities that are "created" by μ^+ and their structures and reactivities are "probed" by μ^+ itself, they are not dealt with here in detail, since they have been fully reviewed [12, 49].

Although the high-field μSR method has provided data for various Mu-substituted free radicals, it can only determine the muon hyperfine constants because the system is decoupled and the muon spin rotation is insensitive to the nuclear hyperfine interaction. The possibility of using LCR was first suggested in 1984 [36] and it is now used widely.

The resonance methods were used very early to determine the fundamental properties of muon and muonium; i.e., μ^+ spin resonance to determine the μ^+ magnetic moment and Mu spin resonance to determine the hyperfine coupling constant of Mu. But chemical applications of Mu resonance methods were scarce. In the one reported application of Mu resonance, Mu was observed in Ar gas and Mu reactions with O_2, NO and C_2H_2 were measured from the change in the resonance single [50]. Since an extremely high RF power is required to induce H_1 so that the μ^+ spin on-resonance evolves at least a few times within the time window of observation, resonance methods have not been used widely at muon factories where only the dc mode μ^+ beam is available. Application of the μ^+ resonance method to chemistry and materials science was first done at UTMSL (University of Tokyo, Meson Science Laboratory) taking advantage of its pulsed muon beam, since a high RF power could be applied in a pulsed mode synchronized to the muon pulses with a low duty factor [34]. The first μ^+ spin resonance apparatus consisted of a 3 kG magnetic field and 40 MHz RF power. This was scaled up to a system in which a 10 kW RF power of 500 MHz was fed into a gigantic cavity (80 cm in diameter) and a magnetic field of up to 37 kG which was supplied by a superconducting solenoid coil. This has been applied to various studies.

The mechanism of Mu formation is a matter of substantial importance. The scheme illustrated in Fig. 2 assumes that Mu is formed at the end of the charge-exchange cycles, and forms the basis of the "epithermal model" of Mu formation. However there is another view, a "spur reaction model" of Mu formation, in which it is assumed that μ^+ stops within or near the terminal spur of the muon track where it can combine with an e^- to form Mu. This model comes from an analogy with the same model for positronium (Ps) formation. Although the spur reaction model of Ps formation has received substantial support for condensed phases [51], the validity of the same model for Mu formation is still open to debate [52].

5 Muonium as Hydrogen-like Probes

Mu is used as hydrogen-like probes to study the atomic and radical reactions of H because Mu reaction rates are directly observable by the Mu decay kinetics method (Eq. 17) or the residual polarization method, while most H data come from measured relative rates. Although the H atom is the simplest and one of the most important species in chemistry, studies of H atoms are very limited. The difficulties lie in the ways of producing H and in its detection. Methods for producing H atom in condensed phases are mainly by radiolysis, but here the chemistry is complicated because a large number of accompanying transients, e_{aq}^-, $\dot{O}H$, H_{aq}^+, H_2 and H_2O_2, are produced. Photolysis is another approach to H atom production, but its applicability is limited. From the side of detection, optical detection of H is not common because of its very small absorption coefficient, and ESR is used more frequently. Although ESR is powerful, its disadvantage is that it has a slow time resolution and thus even when H atoms are observed and their decay is monitored, it is usually concerned with trapped H. Compared to these difficulties of H atom chemistry, Mu has evidently significant advantages since it can be observed directly on the μs time scale. Substantial data on Mu reaction rate constants have been accumulated and some discussions have been made for comparisons between k_M and k_H in the gas phase [22, 53] and in aqueous solutions [43, 54]. In this section, some examples are given to see how Mu can be used as H-like probes.

5.1 Initiation Kinetics of Radical Polymerization

Both H and Mu can attach to unsaturated bonds. If H attaches to the unsaturated bond of monomer molecules, monomer free radicals are formed. The free radicals can initiate polymerization when the next monomer molecule adds to them, and by successive addition of monomer molecules, the polymerization is propagated.

$$H + M \rightarrow H\dot{M} \tag{28}$$

$$H\dot{M} + M \rightarrow HM\dot{M} \ ... \rightarrow HM^n \ . \tag{28'}$$

The rate of the H-initiation reaction 28 is known for a limited number of monomers. In general Mu and H reactions are known to have large isotope effects; e.g. k_M/k_H is ≈ 0.01 for hydrogen abstraction reactions, ≥ 2 for bromine abstraction reactions, ≤ 1 for reduction reactions [43]. However the isotope effect is not so large in addition reactions, and thus use of Mu as a H-like probe for initiation kinetics becomes substantiated [55, 56]. The rate constants of the Mu addition reaction were measured directly by Mu decay kinetics for several monomers. The results are summarized in Table 3 in which the corresponding data for H are also shown when available.

Table 3. Reaction rate data of Mu and H addition reaction to monomers

Monomer	$k_M/10^{10}$ $M^{-1} s^{-1}$	$k_M/10^{10}$ $M^{-1} s^{-1}$	Isotope Effect (k_M/k_H)
Styrene	2.0		
Methylmethacrylate (MMA)	0.95 ± .19		
Acrylonitrile (AN)	1.14 ± .20	0.40 ± .05	2.8
Acrylic acid (AA)	1.55 ± .25		
Acrylamide (AAm)	1.90 ± .13	1.8	1.1
Maleic acid (MA)	1.1	0.80	1.4

The data of acrylamide (AAm) and maleic acid show that there is little isotope effect. As these rate constants are near the diffusion controlled limit of 2×10^{10} $M^{-1} s^{-1}$, it indicates that the diffusion of small particles like H and Mu may be controlled by the radius and not by the mass.

The differences in the rate constants k_M may be explained through resonance and steric effects. Since styrene is the most highly conjugated of the monomers studied, it is not surprising that it has the largest rate constant. The rate constants for the other monomers increase in the order MMA < AN < AA < AAm. MMA has the least accessible double bonds due to the "blocking" effect of the α-methyl groups. The other three are all straight chain three carbon conjugated compounds with the double bonds equally open to Mu attack. Mu, which acts like a nucleophile, would encounter less electron shielding and react faster with acrylamide or acrylic acid, both having carbonyl oxygens to withdraw electron density from the target atoms than it would with acrylonitrile. Although there are no corresponding H data, a similar reactivity trend can be inferred.

In order to prove that Mu is indeed adding across the vinyl double bonds, high-field μSR can be performed for the samples containing neat or high concentration of monomers.

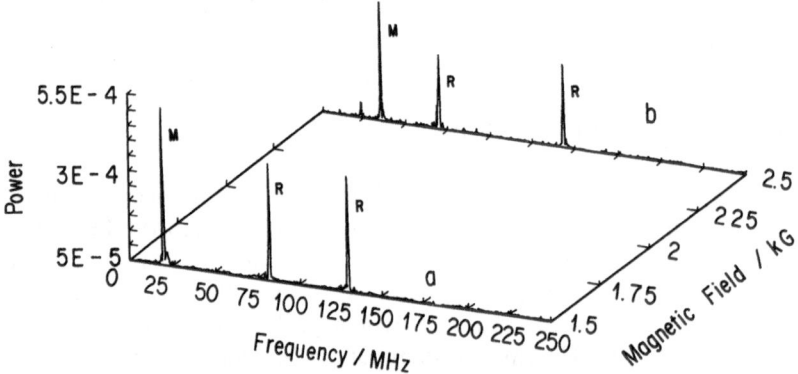

Fig. 8. Fourier transformation spectra of Mu-substituted radicals of (**a**) styrene and (**b**) benzene at 3000 G.
M is the signal of diamagnetic muon (with the frequency ω_μ), and the two peaks (R) are the signals of the Mu-substituted radicals (with hyperfine constant ω_0) corresponding to Eq. 14

Figure 8 shows the results of high-field μ SR for styrene. By Fourier transformation of the measured asymmetry spectra, the frequencies of the radicals and diamagnetic muons are obtained, and the hf frequencies are calculated from Eq. 14. The obtained hf frequencies are listed in Table 4 together with the data of the H analogue. The data of radicals formed by Mu addition to benzene (cyclohexadienyl radical) is also shown for comparison.

If there are no isotope effects in the hf coupling for the Mu- and H-adding monomers, the measured frequency, after compensated for the ratio of proton and muon dipole moments $g_p\mu_p/g_\mu\mu_\mu = 0.3141$, should be the same. Clearly there is an isotope effect in the hf constant which is systematically larger for Mu-radicals. Such an isotope effect has been considered as due to a difference in the conformation of Mu- and H-radicals [46, 57]. The data indicates that Mu is adding preferentially to the vinyl bond of styrene and not to the ring. This differs from the previous observation that H adds to the vinyl bond of styrene 15% of the time

Table 4. Hyperfine coupling constants for the radicals formed by Mu addition to selected monomers

Monomer	MuR		HR
	a_μ (MHz)	$a_\mu \times 0.3141$ (MHz)	a_p (MHz)
Styrene	213.4	67.0	50.1
Methylmethacrylate (MMA)	274.9	86.3	59.7
Acrylonitrile (AN)	280.4	88.1	64.4
Acrylamıde (AAm)	319.3	100.3	70.8
Benzene	514.6	161.6	133.7

and to the ring 85% [58]. In view of the time window of μSR, however, it is conceivable that the observed radicals are the stabilized ones after intramolecular rearrangement and not the primary products.

Further interest invoked during these studies is to follow the chain propagation processes of the monomer radicals (reaction 28'). However this reaction is too slow to be observed within the μSR time window. Sections 6-2 and 6-3 describes somewhat related topics.

5.2 Reaction with Thiosulphate

The reaction of Mu with thiosulphate;

$$Mu + S_2O_3^{2-} \rightarrow ? \tag{29}$$

was studied using Mu decay kinetics and residual polarization methods [59]. The determination of the reaction rate constant was straightforward by Mu decay

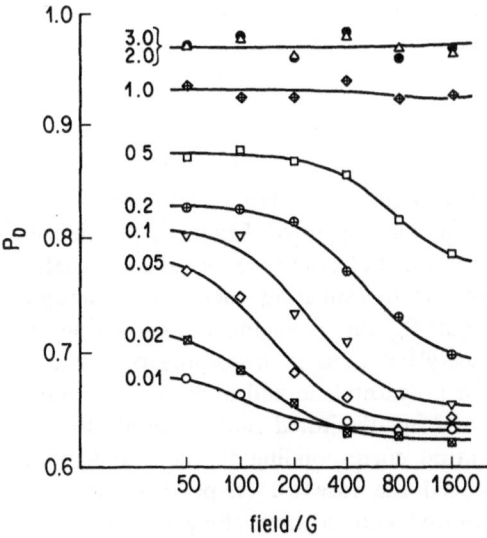

Fig. 9. The results of the residual polarization experiments for the reacting system of Mu with thiosulphate (reaction 29) in aqueous solution. P_D is the diamagnetic polarization which is the sum of the diamagnetic polarization in neat solvent and that formed on reaction 29. The number at left of each curve is the concentration of thiosulphate in mol/dm³. From these results, using Eq. 18 and theoretical residual polarization as in Fig. 5-A, the reaction rate constant is derived.

kinetics in dilute $S_2O_3^{-2}$ solutions, and the result was $k_M = (1.5 \pm 4) \times 10^{10} \, M^{-1} \, s^{-1}$. The residual polarization method was applied for concentrated $S_2O_3^{2-}$ solutions as a function of magnetic field. These results are shown in Fig. 9.

The solid curves are the result of theoretical fitting made under the assumption that the product of reaction 29 is the diamagnetic muon. The fitting parameters were: h = 0.63–0.65 and log k_M = 10.34. The rate constant obtained from the residual polarization method was found to be a little larger than that obtained from the direct Mu decay kinetics. This difference may originate from the difference of the solute concentration and hence of the time scale under investigation; in Mu decay measurements very dilute solute must be used and the rate constant corresponds to a sufficiently slow time scale. In the residual polarization measurements, high solute concentrations are used and eventually the time scale is much faster at which the time-dependent rate constant [60] has to be considered.

The shapes of the field dependence (Fig. 9) suggest that it is Mu that is reacting, and this is confirmed by the Mu decay kinetics. A question that may arise is whether the diamagnetic muon is the direct product or not. The diamagnetic muon may be produced directly as,

$$Mu + S_2O_3^{2-} \rightarrow MuS^- + SO_3^- . \tag{29-2}$$

Another possibility is that the direct product is the radical,

$$Mu + S_2O_3^{2-} \rightarrow Mu\dot{S} + SO_3^{2-} \tag{29-3}$$

but it subsequently reacts and becomes a diamagnetic muon as,

$$Mu\dot{S} + S_2O_3^{2-} \rightarrow MuS^- + S_2O_3^- . \tag{29-3}$$

The results of the residual polarization seem to support the assumption that the product of reaction 29 is the diamagnetic muon, and this implies that, even if these reactions schemes are really taking place, reaction 29-3 should proceed quicky so that no significant loss of polarization occurs in the step.

5.3 Solitons in Polyacetylene

trans-Polyacetylene has attracted special interest since occurrence of soliton, a one-dimensional diffusion of an unpaired electron, was suggested. Magnetic resonances have been mainly used for the study of soliton, but studies initiated at UTMSL have proved that μ^+ is a powerful tool for investigating them in its unique way [61]. The remarkable observation was that P_D, the diamagnetic polarization, of µS Rotation was far smaller in *cis*-polyacetylene than in *trans*-polyacetylene. It is natural to consider that Mu is formed in polyacetylene and adds to the double bonds. The result may be the formation of Mu-substituted radicals as illustrated in Fig. 10c and consequently the polarization corresponding to the diamagnetic muon should become smaller. This is indeed the case for *cis*-polyacetylene. If, however, the unpaired electron is not localized and diffuses along the chain (as soliton in *trans*-polyacetylene), P_D may apparently become larger. This can be easily understood if an extreme case is imagined where the unpaired electron is quickly removed leaving μ^+ in the diamagnetic environment.

The results of μS Relaxation and μS Repolarization carried out for the two isomers are shown in Fig. 10a and b. For *cis*-polyacetylene the asymmetry simply rises with the magnetic field intensity. From the dependence of the asymmetry on the magnetic field, the hf coupling μS constant is estimated to be around 200 MHz for the free radicals (Eq. 24). For *trans*-polyacetylene the initial polarization also rises with the magnetic field. A significant difference is that relaxation is observed, and the rate of the relaxation is field dependent. Such relaxation cannot be due to dipolar moments of neighboring protons since a similar relaxation takes place for D-substituted *trans*-polyacetylene. It is due to spin-lattice relaxation induced by the dipolar hf coupling interaction with the unpaired electron randomly walking along the polymer chain

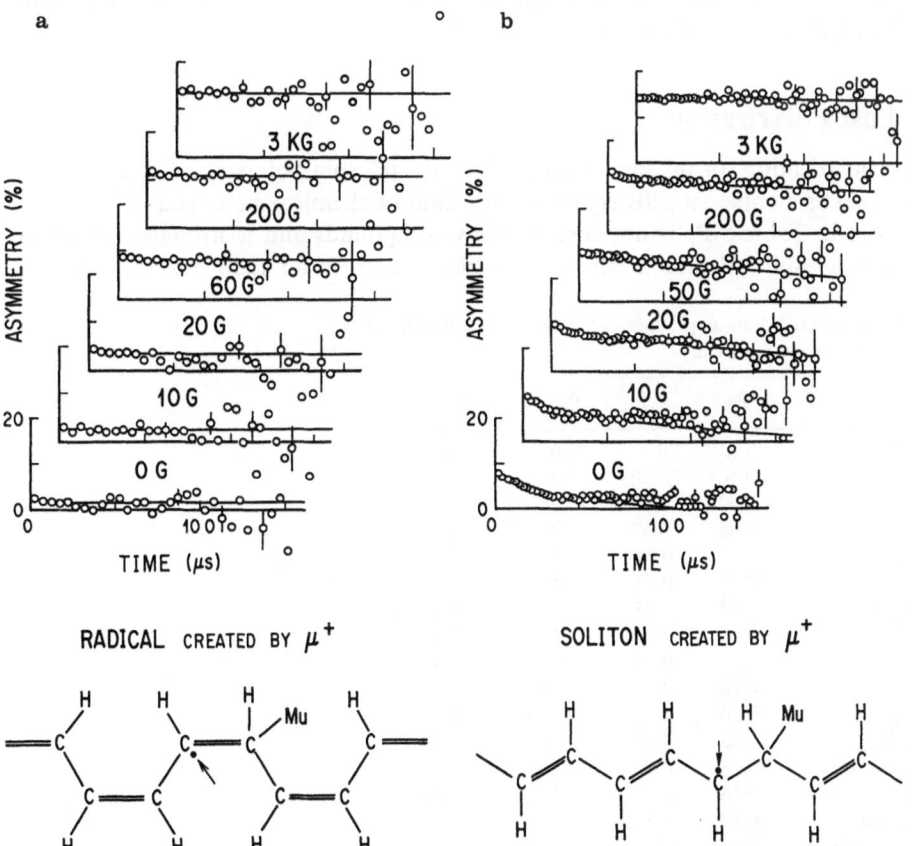

Fig. 10a, b. μS Relaxation and μS Repolarization of polyacetylene. (**a**) Mu addition to *cis*-polyacetylene leads to Mu-substituted radicals formation. Consequently the asymmetry is increased (Repolarization) at higher magnetic field intensity, but there is no relaxation of μ^+ spin. (**b**) Mu adds to *trans*-acetylene, but there are no stable Mu-substituted radicals because the unpaired electron goes into a soliton movement. μ^+ spin is repolarized at high magnetic fields, but there is an apparent relaxation caused by the random walk of the unpaired electron (soliton). The relaxation time due to the soliton created by μ^+ is approximately 100 times smaller than that observed by NMR for the soliton induced by implanted paramagnetic centers. The factor of about 10 may come from the difference of the dipole moments $(g_\mu \mu_\mu / g_p \mu_p)^2 = (1/0.3141)^2$, and the further factor of 10 should correspond to the ratio of the local density of soliton; roughly speaking one μ^+ creates one soliton which is monitored by μ^+ itself, while the density of the solitons created by paramagnetic centers divided by the monitor density (protons) is about ten times smaller. The data of Fig. 10 do not have enough statistics. The data taken with better statistics thereafter indicate that the relaxation function is not a simple exponential decay, and it levels off at a certain asymmetry value depending on the field intensity [62]. This means that the unpaired electron moving as soliton has a probability of jumping to a neighboring chain, and at that moment it is removed from interactions with the μ^+ spin. Thus while the soliton diffuses along the chain, it can also jump to another chain. The intra-chain diffusion is almost temperature independent, while the interchain spin hopping was slightly temperature dependent, indicating that it is the process which requires activation.

6 Lost Polarization

Lost polarization (or missing fraction) P_L is the μ^+ spin polarization that cannot be observed as a coherent µSR signal for any kind of chemical states (Eq. 16).
As seen in Table 5 P_L is not zero in many compounds and it strongly depends on the phase. Spin relaxation which takes place much faster than the resolving time

Table 5. Selected data on μ^+ spin polarizations (Liquid phase unless otherwise specified.)

substance	P_D	P_M	P_R	P_L
CCl_4	1.0	0	0	0
H_2O (gas)	0.10	0.90	0	0
(liq.)	0.62	0.20	0	0.18
(solid)*	0.48	0.52	0	0
D_2O (liq.)	0.57	0.18	0	0.25
(solid)	0.39	0.63	0	0
CH_3OH	0.62	0.23	0	0.15
$n-C_6H_{14}$	0.65	0.13	0	0.22
$c-C_6H_{12}$	0.69	0.20	0	0.11
C_6H_6	0.16	0	~0.63	~0.21
C_6F_6	0.20	0	0.35	0.45
$Si(CH_3)_4$	0.63	0.18	—	0.19
$(CH_3)_2CO$	0.54	0	0.43	0.03
CS_2	0.16	0	?	~0.84
CH_4 (gas)	0.12	0.8	0	0
He (gas/liq.)	1.0	0	0	0
Ar (liq.)	0.02	0.48	0	0.5
Ar (solid)	0.01	0.91	0	0
Kr (gas)	0	1.0	0	0
Kr (liq.)	0.07	0.57	—	0.36

* At temperatures below 200 K. Above 200 K P_M decreases accompanied by an appearance of P_L.

of μSR can cause the loss of polarization. As has been explained, hyperfine relaxation of ^{S}Mu takes place within 0.22 ns, but the loss of ^{S}Mu spin polarization by this process is already taken into account in calculating P_M (Eq. 15). However, if ^{T}Mu is converted to ^{S}Mu by spin-exchange reactions, the latter will become a further source of loss of polarization. Another source of the loss of polarization is the loss of spin coherency. If, for example, there is a reaction by which ^{T}Mu is converted to diamagnetic muon, and if the reaction rate is moderately slow (on the ns time scale) so that the μ^+ spin has time to evolve, μ^+ in the diamagnetic muon becomes out of phase. Although it is believed that both of these processes are the causes of the lost polarization, there has been little done on this problem.

6.1 Lost Polarization in Liquid Water

An explanation for the lost polarization in water was first introduced as part of the spur model of Mu formation and reactions as shown in Fig. 11 [63].
Here Mu is assumed to be formed as a result of combination of μ^+ and an excess electron. This view is the same as for the "spur model" of positronium (Ps) formation. While the "spur model" has received strong support for positronium yield in condensed phases, the validity of the same model for Mu formation is not clear. Figure 11 presents the original form of the spur model of Mu formation, since it helps to contrast the difference between the epithermal model (Fig. 2) and the spur model of Mu formation. Alternatively, the part of Mu formation, i.e., μ^+ and excess electron combination, in Fig. 11 may be replaced with the picture of Mu

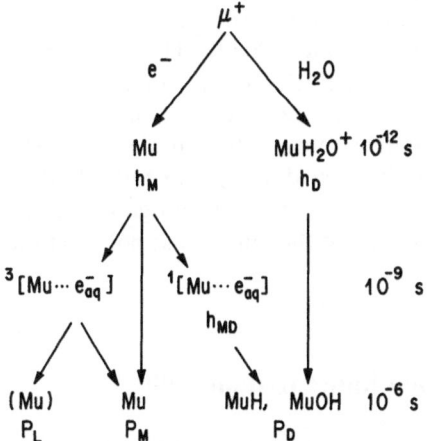

Fig. 11. A proposed scheme of Mu formation and reactions in water. Mu is formed by reaction with nearly thermalized μ^+ and an excess electron formed by radiolysis (the spur model of Mu formation) with the probability h_M, and the rest (h_D) is solvated becomeing diamagnetic muon. Mu thus formed can further react with aqueous electrons; by spin exchange reaction it is depolarized and the polarization is lost (P_L). By chemical reactions, a fraction of Mu is incorporated into diamagnetic muon while another fraction becomes dephased and forms a part of P_L depending on how fast the reactions proceed. Mu that has not reacted is observed as free Mu (P_M)

appearing from the charge-exchange cycles. Other aspects of Fig. 11, i.e. that Mu is necessarily subject to reaction with the transient products created by μ^+ radiolysis, may be acceptable. If the nascent Mu, formed with initial fraction h_m, survives chemical reactions, it is observed as coherent Mu precession and is counted in P_M. If it is depolarized by spin-exchange reaction with any species having an unpaired electron as with the aqueous electron, e_{aq}^-, then the lost polarization occurs.

$$^TMu + e_{aq}^- \rightarrow {}^SMu + e_{aq}^- . \tag{30}$$

Although P_L was initially attributed to spin-exchange reaction, formation of dephased diamagnetic muon (MuH below) as a result of reaction

$$^TMu + e_{aq}^- \rightarrow MuH + OH^- \tag{31}$$

cannot be excluded, as invoked by the already well known reaction for H atom,

$$H + e_{aq}^- \rightarrow H_2 + OH^- . \tag{32}$$

It is fortunate that μSR provides a means to distinguish between the depolarization by spin-exchange and dephasing by slow Mu reaction. Using the residual polarization method, it is possible to determine the Mu fraction h_{MD} at the moment of reaction 31. The experiment gave $h_{MD} = 0.08$ for the Mu fraction, and $\lambda_{31} = 6 \times 10^8$ s^{-1} for the reaction rate at room temperature [64]. The polarization corresponding to the spin-exchange is therefore $1 - h_D - h_M - h_{MD} = 0.10$. Thus ^TMu can undergo both chemical reactions and spin exchange reactions on encountering aqueous electrons, with both resulting in lost polarization. The value of P_L and its partition between reactions 30 and 31 are found to be temperature dependent.

The lost polarization in water thus has two origins, but both are different reaction routes for Mu encounter with E_{aq}^-. In ice there is also lost polarization, but it gradually decreases at lower temperatures and disappears below 200 K. The decrease of P_L in ice is accompanied by an increase in P_M. This is attributed to temperature dependent diffusion rate of Mu and H atoms which, on encounter, can exchange electron spins [63b]. Similar processes may account for the lost polarization in organic solids and liquids. Although it can be supposed that it is H atom or free radicals that encounter and react with Mu in hydrocarbons [33, 65], detailed investigations have not yet been performed. Mu-substituted radicals are also thought to lose polarization in this way [66, 67].

6.2 Measurements of Free Radicals Reaction Rate Constants [68]

In this section, measurements of the reaction rate constants of Mu-substituted radicals are illustrated taking Mu-substituted cyclohexadienyl radicals (Mu$\dot{C}_6H_6$) as an example. The subject not only shows the difference of the μS Rotation and μS Resonance methods but also serves as an introduction to the next section.

The reaction studied is,

$$Mu\dot{C}_6H_6 + C_6H_4O_2 \rightarrow \text{diamagnetic muon?} \tag{33}$$

Mu-substituted hexadienyl radicals are easily observed by high-field μS Rotation and their reaction with benzoquinone has been studied.[47] The method is principally the same as in the Mu relaxation kinetics but, since Mu-substituted radicals are measured, the disappearance rate of $Mu\dot{C}_6H_6$ is determined from the line broadening of the Fourier transformation power spectrum. The obtained value is $k_{33} = 2.8 \times 10^8$ M^{-1} s^{-1}. It is important to note that the reaction rate constant of Mu-substituted radicals may not differ much from that of the H-counterparts. The isotope effect should be small since the mass of Mu- and H-radicals are not much different, and since the reaction center is separated from the site of Mu- and H-addition.

The direct product of reaction 33 is not known. Inferring from the reaction data of dihydro-cyclohexadienyl radicals with benzoquinone, the product is thought to be diamagnetic [69]. In the Rotation method μ^+ spin in the products of reaction 33 is dephased, and it is impossible to know whether it is diamagnetic or a radical. The same reaction has been studied by μ S Resonance [70]. Here the magnetic field was set at the resonance condition of $MuC_6\dot{H}_6$, and the reaction rate constant

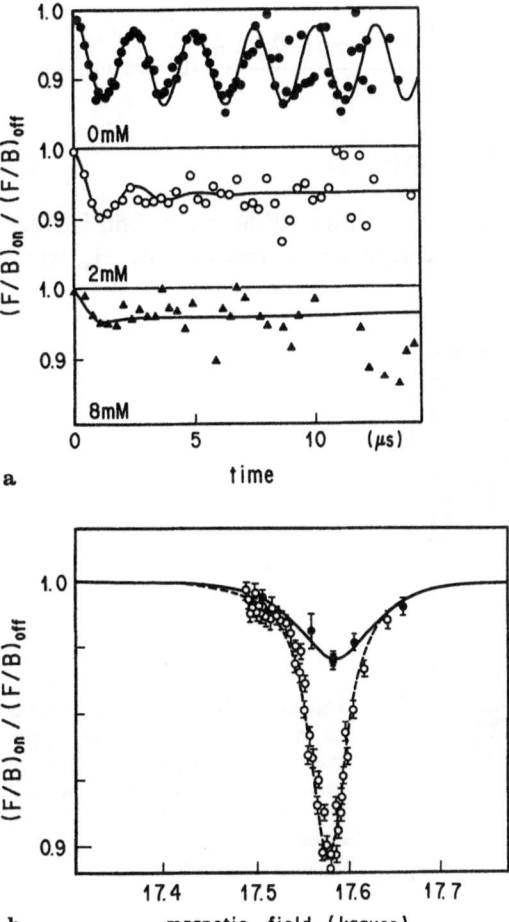

Fig. 12a, b. μS Resonance of benzoquinone solution in benzene. a: The spin evolution at just-resonance of the Mu-substituted radicals of benzene (Mu-substituted hexadienyl radicals) in neat benzene (upper spectrum). When benzoquinone is added (lower two spectra, 2 mM and 8 mM), the spin evolution is damped showing that Mu-substituted hexadienyl radicals are reacting with benzoquinone. The reaction rate constant is derived from this damping. b: The Mu-substituted hexadienyl radicals resonance in neat benzene presented by the resonance lineshape (open circles). When 8 mM benzoquinone is added (black circles), the resonance signal becomes weak and the linewidth is broadened. The reaction rate constant is derived from the broadening of the linewidth, too

121

was determined both from the relaxation rate of the spin evolution time histogram and from the line broadening of the resonance curve.

The value are 2.6×10^8 and 3.0×10^8 M^{-1} s^{-1}, respectively, and they agree well with the data from the Rotation method. If the product of reaction 33 is the diamagnetic muon, its formation rate must be the same as the reaction rate of $MuC_6\dot{H}_6$. The formation rate of the diamagnetic muon was monitored with the same resonance apparatus by switching the magnetic field to the resonant condition of the diamagnetic muon. These results are shown in Fig. 13. Compared with the simple spin evolution as in Fig. 7, a peculiar shift of the baseline is noted.

The baseline shift shows how the Mu-substituted hexadienyl radicals are converted to diamagnetic muons by reaction 33, and start to evolve as a member of the resonant component. The following equation describes this situation, where the first term is the evolution of the diamagnetic muon already present at $t = 0$ (with the asymmetry A_0^{res}), the second term is that produced by reaction 33 with the rate $\lambda_{33} = k_{33}$ [Quinone] at time $t = t'$ and starts to evolve thereafter, and the third term is the asymmetry of Mu-hexadienyl radicals which are not on-resonant and are decreasing due to the reaction.

$$R(t) = A_0^{res} \cos(\omega_{res}t) + \int_0^t \frac{d[A_0^{non-res}(1 - \exp(-\lambda_{33}t'))]}{dt'} dt' \cos(\omega_{res}(t - t'))$$

$$+ A_0^{non-res} \exp(-\lambda_{33}) \qquad (34)$$

The fitting of this equation to Fig. 13 gives $k_{33} = 0.56 \times 10^8$ M^{-1} s^{-1}. This is much smaller than 2.8×10^8 determined from the disappearance rate of MuC_6H_6, which

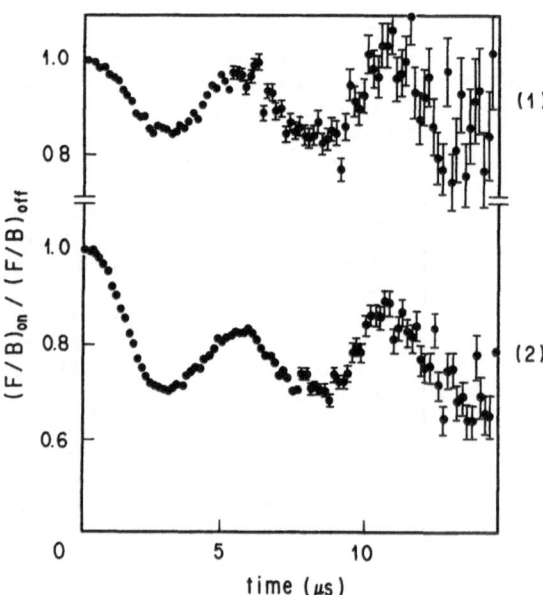

Fig. 13. μ S Resonance of diamagnetic muon for neat benzene (upper) and 8 mM benzoquinone solution of benzene (lower spectrum). The baseline shift of the lower spectrum indicates that new diamagnetic muon is being produced slowly

means that the diamagnetic muon is not the direct product of reaction 33 and that there are some further intermediate reaction steps before the diamagnetic state is formed.

6.3 Lost Polarization in Carbon disulfide [68]

Of all the chemical species for which μSR were measured, CS_2 has the largest lost polarization of 84% (Table 5). No Mu has been observed in it, but this is not surprising because CS_2 has unsaturated bonds and Mu, if formed, may add to them. In such a case Mu-substituted radicals are expected instead. It was a good challenge to search for Mu-substituted radicals in CS_2 at KEK BOOM where a powerful μS Resonance spectrometer was operating. In principle, Mu-substituted radicals can be measured more easily by high-field μS Rotation than by μS Resonance. In the Rotation method all the μ^+ spins evolve simultaneously and reveal themselves in the Fourier transformation power spectrum at their intrinsic frequencies. In the Resonance method, on the other hand, the RF frequency is fixed and the magnetic field must be swept-by-step to see whether the μ^+ spin asymmetry has

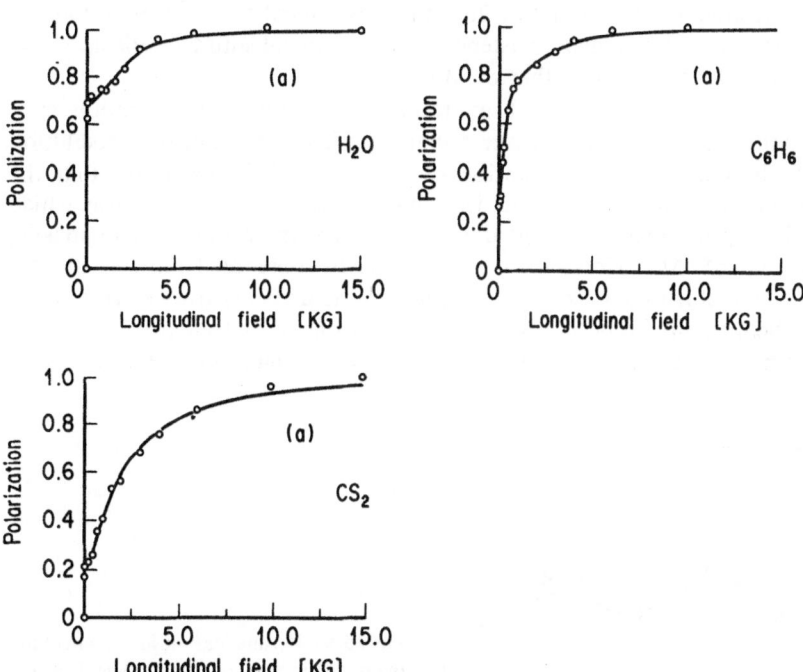

Fig. 14. μS Repolarization for H_2O, C_6H_6 and CS_2. The steepness of the rise of polarization contain information about the species involved in the loss of polarization at low fields. Mu and Mu-substituted hexadienyl radials are responsible for the loss of polarization in H_2 and C_6H_6, respectively. The repolarization curve of CS_2 suggests that a species with a hyperfine constant between those of Mu and Mu-substituted hexadienyl radicals is the precursor of the lost polarization

dropped by spin evolution on-resonance. This is an excessive time consuming task, and no one would dare it unless being compelled by some special necessity. And CS_2 presents this very special situation; i.e. no Mu-substituted radicals have been found by high-field μS Rotation. It was thought that Mu-substituted radicals must have depolarized or dephased, and μS Resonance should become a powerful diagnostic tool for such a case.

Before attempting experiments of μS Resonance, μS Repolarization was performed [33]. The results are given in Fig. 14 in which the data for H_2O and C_6H_6 are shown for comparison. In all the data the polarization at zero field are equal to the diamagnetic muon polarization P_D. This is because the field was not completely zero: the sample position was affected by small fringing fields from the magnets of the beam-transport system and the earth's magnetic field. This resulted in spin evolution of Mu and Mu-substituted radicals for which the gyromagnetic ratio is larger than that of diamagnetic muons. As has been explained in the previous section, the precursor of the loss of spin polarization in water is Mu, and the shape of the repolarization curve is fitted by Eq. 25 and 26 with ω_0 equal to the hf frequency of Mu. For C_6H_6 the repolarization shape is much steeper. Equation 26 could be fitted with $\omega_0/2\pi = 514$ MHz which is the hf frequency of Mu-substituted radicals in benzene ($Mu\dot{C}_6H_6$). The repolarization in CS_2 is steeper compared to that in water, but is moderate compared to that in benzene. The hf frequency can not be determined in a unique way from Eq. 26, but is estimated to be around 1000 to 2000 MHz. The results thus suggest existence of some Mu-substituted radicals whose hf coupling is a little smaller than that of Mu.

No Mu-substituted radicals have been found by high-field μS Rotation even with the specially arranged high transverse magnetic field and high time resolution that enabled observation of the Rotation of up to 1 GHz. This meant that the loss of polarization is taking place in less than 1 ns. The μS Resonance which uses a high decoupling magnetic field does not involve the problem of dephasing and hf relaxation of ^SMu. However a search for Mu-substituted radicals in neat CS_2 was not successful in spite of the persistent measurements over a wide range of magnetic fields [70]. Only in dilute CS_2 solutions in tetrahydrofurane was some symptom of radicals, but its asymmetry was small and readily disap-

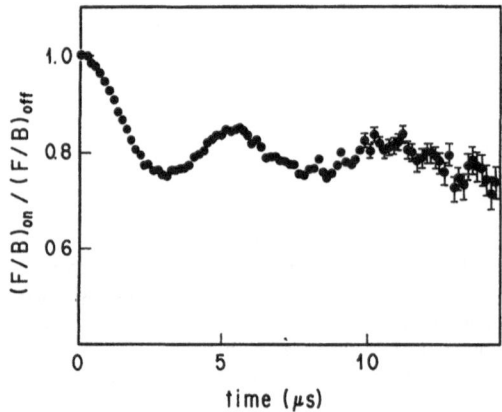

Fig. 15. μ S Resonance time histogram of the diamagnetic muon in CS_2. The baseline shift indicates that new diamagnetic muons are being formed, and the slight relaxation shows that the diamagnetic muons are disappearing gradually

peared when the CS_2 concentration was increased. The hf coupling constant of this symptomatic component was small, even smaller than the value (2.8 MHz) reported as being Mu-substituted radicals of CS_2 in dimethylbutadiene.[71] These small hf constants have not yet been assigned.

Apart from the radicals search, the Resonance of the diamagnetic muon in CS_2 was also performed using the same apparatus. Figure 15 shows the results, which are very similar to Fig. 12; clearly the asymmetry on-resonance is damping and the baseline is shifting. This was analyzed using an equation similar to Eq. 34, but including a relaxation function in the first term. This means that the diamagnetic muon is gradually formed on the time scale of μs and at the same time disappearing. The shift of the baseline is temperature dependent and the Arrhenius plot of the appearance rate results in an activation energy of 1.5 kcal/mol. This value is considerably small for ordinary chemical reaction.

Summarizing, no stable μ^+ containing species, except for the already recognized diamagnetic muon (16%) have been found by using all the techniques of μSR. No Mu and no Mu-substituted radicals have been observed, and the diamagnetic muon itself seems to be in a dynamic change. Combining all these observations, the muonic processes as shown in Table 6 are proposed. Mu and μ^+ are formed on muon injection into CS_2. This initial μ^+ polarization may correspond to $P_D = 0.16$ observed by μS Rotation. CS_2 is known to form an anion by radiolysis [70], and ^TMu can exchange its electron spin and be converted to ^SMu. ^SMu will be depolarized at low fields, and this is the main pathway for the lost polarization. When a decoupling high magnetic field is applied, the spin depolarization due to ^SMu hf relaxation should become negligibly small. Even under such conditions, however, there are the reactions (d) to (g) that bring about the μ^+ spin from one chemical state to another. Reaction (d) is the Mu addition to CS_2. Since no Mu-substituted radicals have been observed, $Mu\dot{C}S_2$ is supposed to be short-lived in neat CS_2 and will add the second CS_2 molecule (reaction (e)) in less than 1 ns. In polymerization of vinyl-monomers, the second addition reaction like reaction (e) is slow (of the order 10 μs or longer), but, as the small activation energy of the appearance rate indicates, reaction (e) for CS_2 may not be the same as the polymerization and is rather a kind of clustering process. A Mu-containing aggregated state may develop by adding further molecules (reaction (f)), and ultimately the hf coupling becomes negligibly small because the μ^+ and e^- become separated (reaction (g)). Such is in effect the diamagnetic muon, and it explains the gradual growth of the diamagnetic muon signal. Reaction (h) is suggested to explain the slow decay of the diamagnetic muon.

Table 6. Proposed muonic processes in liquid CS_2

(muon injection)	$\rightarrow$ $^{T/S}Mu$, μ^+	(a)
$^TMu + (CS_2)^-$	$\rightarrow$ $^SMu + (CS_2)^-$	(b)
^SMu	$\rightarrow$ depolarize at low fields	(c)
$Mu + CS_2$	$\rightarrow$ $Mu\dot{C}S_2$	(d)
$Mu\dot{C}S_2 + CS_2$	$\rightarrow$ $Mu(\dot{C}S_2)_2$	(e)
$Mu(\dot{C}S_2)_2 + CS_2$	$\rightarrow$ ... $\rightarrow Mu(\dot{C}S_2)_n$	(f)
	$\rightarrow$ μ^+ ... CS_2^-	(g)
μ^+ ... CS_2^-	$\rightarrow$ $Mu + CS_2$	(h)

The scheme of Table 6 has yet to be confirmed by further experiments, not only of μ SR itself, but of other fields like radiation chemistry. There are few studies of CS_2 radiation chemistry, but a recent study using pulse radiolysis showed results that parallel the muon chemistry [70].

7 Concluding Remarks

This review has been aimed at guiding the readers toward the use of positive muons in chemistry emphasizing the special aspects, i.e. the processes which are "induced and probed" by positive muons, and illustrating examples of all the techniques of muon science, except that of level crossing resonance. No attempt has been made to refer to all important works; for example, an omitted topic is μ SR studies of crystalline metal complexes [72]. Here μ^+ appears to take a position, slowly hopping, about 0.2 nm away from the central metal cation. It would be interesting to examine whether it is a new chemical state "induced and probed" by μ^+, or is the one also formed when p^+ is injected into the same substance. Another omitted example is the phase dependence of diamagnetic muon polarization, P_D, in various substances: These values are known to change for gas, liquid and solid phases [73]. The nature of this phase dependence is not understood yet, it is undoubtedly the outcome of chemical states induced and probed by μ^+. There are many more interesting subjects in muon chemistry to be viewed from the standpoint of this article, which, by interplay with hot atom chemistry and radiation chemistry, will contribute substantially to basic understanding and applications.

8 Acknowledgements

Thanks are due to the author's colleagues, Prof. Y. Tabata, Drs. Y. Miyake and T. Azuma, and to the members of UTMSL, Prof. K. Nagamine and Dr. K. Nishiyama. Gratitude is also extended to the colleagues at TRIUMF, Prof. D. C. Walker, and Drs. B. W. Ng and J. M. Stadlbauer. Comments of Prof. Walker to the manuscript are highly appreciated.

9 References

1. Hughes VW (1966) Ann. Rev. Nucl. Sci., vol 16, p 445
2. Goldanskii VI, Firsov VG (1971) Ann. Rev. Phys. Chem., vol 22, p 209
3. Brewer JH, Crowe KM, Gygax FN, Schenck A (1975) Muon physics, vol 3, Academic Press, p 3
4. Schenck A (1976) Nuclear and particle physics at intermediate energies, Plenum Press, p 159
5. Fleming DG, Garner DM, Vaz LC, Walker DC, Brewer JH, Crowe KM (1979) Positronium and Muonium Chemistry: Adv. Chem. Ser., vol 175; p 279
6. Walker DC (1983) In: Muon and muonium chemistry, Cambridge Univ. Pres.
7. Chappert J, Grynszpan RI (eds) (1984) Muons and pions in materials research, North-Holland
8. Schenck A (1985) In: Muon spin rotation spectroscopy, Adam Hilger, Bristol
9. Schneuwly H, Pokrovsky VI, Ponomarev VI (1978) Nucl. Phys. A312: 419

10. Fleming DG (Private communication)
11. Roduner E, Fischer H (1981) Chem. Phys. 54: 261
12. Roduner E (1984) In: Chappert J,Grynszpan RI (eds) Muons and pions in materials research, North-Holland, p 209
13. Roduner E, Brinkman GA, Louwrier WF (1982) Chem. Phys. 73: 117
14. Roduner E (1986) Radiat. Phys. Chem. 28: 75
15. Jean YC, Brewer JH, Fleming DG, Walker DC (1978) Chem. Phys. Lett. 60: 125
16. Jean YC, Fleming DG, Ng BW, Walker DC (1979) Chem. Phys. Lett., 66: 187
17. Stadlbauer JM, Ng BW, Jean YC, Walker DC (1983) J. Am. Chem. Soc. 105: 752
18. Percival PW, Fischer H (1976) Chem. Phys. 16: 89
19. Percival PW, Newman KE, Spencer DP (1982) Chem. Phys. Lett. 93: 366
20. Arsenau DJ, Garner DM, Senba M, Fleming DG (1984) J. Phys. Chem. 88: 3688
21. Percival PW, Brodovitch JC, Newman KE (1982) Chem. Phys. Lett. 91: 1
22. Fleming DG, Brewer JH, Garner DM, Crowe KM (1981) J. Chem. Phys. 64: 1281 (and references cited therein)
23. Percival PW (1979) Radiochimica Acta, 26: 1
24. Garner DM, Fleming DG, Brewer JH (1978) Chem. Phys. Lett. 55: 163
25. Jean YC, Brewer JH, Fleming DG, Walker DC (1978) Chem. Phys. Lett. 57: 293
26. Ng BW, Jean YC, Ito Y, Walker DC (1981) J. Phys. Chem. 85: 454
27. Stadlbauer JM, Ng BW, Jean YC, Walker DC (1983) J. Phys. Chem. 87: 841
28. Jean YC, Ng BW, Stadlbauer JM, Walker DC (1981) J. Chem. Phys. 75: 2879
29. Fleming DG, Garner DM, Brewer JH, Bowen KM (1977) Chem. Phys. Lett. 48: 393
30. Brewer JH, Crowe KM, Gygar FN, Fleming DG (1982) Phys. Rev., A9: 495
31. Roduner E (1986) Radiat. Phys. Chem. 28: 75
32. Ivanter IG, Minaichev EV, et al. (1972) Soviet Phys. JETP 27: 301
33. Miyake Y, Ito Y, Nishiyama K, Nagamine K, Tabata Y (1986) Radiat. Phys. Chem. 28: 99
34. Kitaoka K, Takigawa M, Yasuoka M, Itoh M, Takagi S, Kuno Y, Nishiyama K, Hayano R, Uemura Y, Imazato J, Nakayama H, Nagamine K, Yamazaki T (1982) Hyperfine Interactions 12: 51
35. Morozumi Y, Nishiyama K, Nagamine K (1986) Phys. Letters A, 118: 93
36. Abragam CR (1984) Acad. Sci. Ser. II 299: 95
37. Garwin RL, Lederman LM, Weinrich M (1957) Phys. Rev. 105: 1415
38. Mobley RM, Amato JJ, Hughes VW, Rothberg JE, Thompson PA (1967) J. Chem. Phys. 47: 3074
39. Percival PW, Fischer H, Camani M, Gygax FN, Rueff W, Schenck A, Scilling H, Graf H (1976) Chem. Phys. Lett. 39: 333
40. Percival PW, Roduner E, Fischer H (1978) Chem. Phys. 32: 353
41. Ito Y, Ng BW, Jean YC, Walker DC (1980) Can. J. Chem., 58: 2395
42. Nagamine K, Nishiyama K, Imazato J, Yoshida M, Sakai Y, Tominaga T (1982) Chem. Phys. Lett., 87: 186
43. Percival PW, Roduner E, Fischer H (1979) In: Ache HJ (ed) Positronium and Muonium Chemistry: Adv. Chem. Ser., 175: 335
44. Venkateswaran K, Barnabas MV, Ng BW, Walker DC (1988) Can. J. Chem. 66: 1979 (and the references cited therein)
45. Roduner E, Percival PW, Fleming DG, Fischer H (1978) Chem. Phys. Lett. 57: 37
46. Roduner E, Strub W, Burkhard P, Hochmann J, Percival, PW and Fischer H (1982) Chem. Phys. 67: 275
47. Roduner E, Brinkman GA, Louwrier WF (1984) Chem. Phys. 88: 143
48. Geeson DA, Symons MCR, Roduner E, Fischer H, Cox SFJ (1985) Chem. Phys. Lett. 116: 186
49. Roduner E (1986) Progress in Reaction Kinetics 14: 1
50. Mobley RM, Bailey JM, Cleland WE, Hughes VW, Rothberg JE (1966) J. Chem. Phys. 44: 4354
51. Ito Y (1988) In: Schrader DM, Jean YC (eds) Positron and positronium Chemistry, Elsevier, p 120 (Studies in Physics and Theoretical Chemistry Series, vol 57)
52a. Walker DC, Jean YC, Fleming DG (1978) J. Chem. Phys. 70: 4534
  b. Miyake Y, Tabata Y, Ito Y, Ng BW, Stadlbauer JM, Walker DC (1983) 101: 372
  c. Mogensen OE, Percival PW (1986) Radiat. Phys. Chem. 28: 85
53. Fleming DG, Garner DM, Mikula JR (1981) Hyperfine Interactions 8: 337

54. Walker DC (1981) J. Phys. Chem. 85: 3960
55. Stadlbauer JM, Ng BW, Walker DC, Jean YC, Ito Y (1981) Can. J. Chem. 59: 3261
56. Stadlbauer JM, Ng BW, Jean YC, Ito Y, Walker DC (1983) In: Initiation of polymerization, ACS Symp. Ser., vol 212, p 35, Am. Chem. Soc.
57. Roduner E (1980) J. Mol. Structure, 60: 19
58. Swallow AJ (1968) Adv. Chem. Ser., vol 82, p 499, Am. Chem. Soc.
59. Percival PW, Brodovitch JC, Newman KE (1984) Faraday Discuss. Chem. Soc. 78: 315
60. Noyes RM (1961) Prog. React. Kinet., 1: 129
61. Nagamine K, Ishida K, Matsuzaki T, Nishiyama K, Kuno Y, Yamazaki T, Shirakawa H (1984) Phys. Rev. Lett. 53: 1763
62. Ishida K, Nagamine K, Matsuzaki T, Kuno Y, Yamazaki T, Torikai E, Shirakawa H, Brewer JH (1985) Phys. Rev. Lett. 55: 2009
63a. Percival PW, Roduner E, Fischer H (1978) Chem. Phys. 32: 353
   b. Percival PW, Adamson-Sharpe KM, Brodovitch JC, Leung SK and Newman KE (1985) Chem. Phys. 95: 321
64. Leung SK, Brodovitch JC, Percival PW, Yu D, Newman KE (1988) Chem. Phys. 121: 393
65. Ito Y, Miyake Y, Tabata Y, Nishiyama K, Nagamine K (1982) Chem. Phys. Lett. 93: 361
66. Roduner E, Brinckman GA, Louwrier WF (1982) Chem. Phys. 73: 117
67. Roduner E, Webster BC (1983) J. Chem. Soc. Faraday Trans. 179: 1939
68. Azuma T, Washio M, Tabata Y, Ito Y, Nishiyama K, Miyake Y, Nagamine K (1989) Radiat. Phys. Chem. 34: 659
69. Raghavan NV, Steenken S (1980) J. Am. Chem. Soc. 102: 3495
70. Azuma T (1987) Doctoral Thesis, University of Tokyo
71. Roduner E (1986) Hyperfine Interactions 32: 741
72. Sakai Y, Kubo MK, Tominaga T, Nishiyama K, Nagamine K (1988) J. Radioanal. Nucl. Chem., Letters 127: 425
73. Ito Y, Miyake Y, Tabata Y, Ng BW, Walker DC (1984) Hyperfine Interactions 17/19: 733

Laser-induced Photoacoustic Spectroscopy for the Speciation of Transuranic Elements in Natural Aquatic Systems

Jae-Il Kim, Reinhard Stumpe and Reinhardt Klenze

Institut für Radiochemie, TU München, 8046 Garching, FRG

Topics in Current Chemistry, Vol. 157
© Springer-Verlag Berlin Heidelberg 1990

Laser-induced Photoacoustic Spectroscopy (LPAS) is a new elegant instrumentation for the chemical speciation of aquatic transuranium (TRU) ions in very dilute concentrations ($>10^{-8}$ mol L^{-1}). The aim of this paper is to review the application of LPAS to the study of the chemical behaviour of TRU ions in natural aquatic systems, the knowledge of which has become increasingly in demand in connection with the safety analysis of nuclear waste disposal in the geosphere. The first part of the paper describes the principle, instrumentation and characteristics of LPAS in aqueous solution, taking particular examples from our own experience. The theoretical estimation of the speciation sensitivity is demonstrated and the result is compared with experiment. The second part deals with the spectral work in aqueous solution and then with the application of LPAS for the speciation of TRU ions in groundwater. Some examples demonstrated are hydrolysis reaction, complexation and colloid generation of the Am^{3+} ion. Speciation sensitivities of U, Np, Pu and Am of different oxidation states in a variety of aqueous solutions are summarized. The present review is so oriented that the expected readers will appreciate the introductory understanding of the LPAS method and its potential applicability in the aquatic chemistry of TRU ions. The application is of course open to a broad field of microchemistry in which the conventional spectrophotometric method has difficulty with sensitivity.

1 Introduction

Growing interest has been recently directed to the application of photoacoustic sensing techniques to the spectroscopic analysis of various optical absorbers in very dilute concentrations. For this purpose a laser is commonly used as a light source. Since the discovery of the photoacoustic effect by A. G. Bell in 1880 [1], its application has a long history of development [2]. Renewed interest in photo-acoustics has emerged starting with the work of Kreuzer in 1971 [3] who analysed trace amounts of gas molecules by laser-induced photoacoustic generation. The theory, instrumentation and application of laser-induced photoacoustic generation developed in recent years have been thoroughly reviewed by Patel and Tam [4] and more recently by Tam [5, 6]. Other reviews are also available in the literature from different authors: Pao [7], Somoano [8], Rosencwaig [9, 10], Colles et al. [11], Kirkbright and Castleden [12], Lyamshev and Sedov [13], Kinney and Stanley [14], West et al. [15] and Zharov [16].

Because of difficulties involved in handling radioactive preparatives, the photo-acoustic sensing technique had not been applied until some years ago to the spectroscopy of aqueous actinide ions. Recently, the authors' laboratory [17–19] has introduced a relatively simple detection apparatus of photoacoustic spectroscopy for the spectral work of actinide ions using pulsed laser as a light source. This detection apparatus can be used for radioactive α-emitting aqueous samples without restriction to corrosive solutions and facilitates the spectroscopic investigation of actinide solutions, particularly transuranic ions, in very dilute concentrations [19]. The spectroscopic system has, since then, been introduced to different nuclear chemical laboratories and further developed for a variety of purposes [20–23]. Most of these developments are confined primarily to the spectroscopic investigation (i.e. speciation) of actinides in very dilute solutions [24–27] or natural aquatic systems [28–34] in which the solubility of actinides is, in general, very low ($< 10^{-6}$ mol L^{-1}). Optical spectroscopy of high sensitivity is an indispensable tool for the study of the chemical behaviour of actinides in natural aquatic systems, which has a newly developing research field in connection with the nuclear waste disposal in the geosphere [28]. For this reason, not only is photoacoustic spectroscopy attracting great attention but also thermal lensing spectroscopy [35–37] and fluorescence spectroscopy [21, 38–41], all using laser light sources, are in growing use for the same purpose.

Actinides have particular spectroscopic properties which are characterized primarily by the $f \rightarrow f$ transitions within the partially filled 5f shell [42] and thus by a number of relatively weak but very sharp absorption bands. The optical spectra of actinides are characteristic for their oxidation states, and to a lesser degree dependent upon the chemical environment of the ion [43]. Thus spectroscopic investigation provides information on the oxidation state of an actinide element [42] and also serves to characterize the chemical states, such as hydrolysis products [44], various complexes [37, 45, 46] and colloids [29, 40]. Hence, laser-induced photo-acoustic spectroscopy (LPAS) with its high sensitivity can be conveniently used for the speciation of aqueous actinides in very dilute concentrations [17–28].

This paper summarizes the present knowledge of laser-induced photoacoustic spectroscopy, as regards theoretical backgrounds, instrumentation and radiochemical applications to particular problems in aquatic actinide chemistry. Since there is no other radiochemical application known in the literature, except the measurement of tritium decay by a acoustic sensing technique [47], the present discussion is limited to application to actinide chemistry, particularly, in aquatic systems. The most interesting field of application is and will be the geochemical study of long-lived radionuclides, namely man-made elements (transuraniums). The main importance for such a study is not only the detection of a migrational quantity of radioactivity but also the characterization of their chemical states and hence their chemical behaviour in a given aquifer systems. Knowledge of this kind will facilitate a better prediction of the environmental impact of transuranic elements which are being produced in ever growing quantities and will be disposed of in the geosphere.

2 Laser-Induced Photoacoustic Spectroscopy (LPAS)

2.1 General Description

The principle of photoacoustic spectroscopy, i.e., the measurement of optical absorption through the detection of acoustic signals produced by the non-radiative relaxation of the excited state, was discovered by Alexander Graham Bell [1]. His experiment was performed with a "Photophone", showing that modulated sunlight which is absorbed by various materials generates acoustic "audible" signals as a function of the wavelength and the strength of the absorption. As early as the following year there were several publications concerning work with a "spectrophone" by Mercadier [48], Tyndall [49], Preece [50], Röntgen [51] and Lord Rayleigh [52]. In the following years, however, interest in the photoacoustic effect subsided. In 1938 [53] the importance of the method for gas analyses was demonstrated. In 1943 [54] there appeared a publication concerning continuous gas analyses in the infrared; using a differential arrangement, a sensitivity of 1 ppm was achieved. The experiment was implemented with a membrane capacitor to detect the pressure variations in the gas due to the absorption of optical radiation.

Since 1971, sensitive microphones coupled to the sample via an optically transparent gas volume and using lasers as powerful light sources have been able to determine spectroscopically small impurities in gases [3, 55]. This sensitive technique has found numerous applications in many fields [56–58]. Papers dealing with the photoacoustic spectroscopy of solutions were first published in 1973 [59, 60]. In analogy with the experiments in the gas phase, gas-coupled microphones were used as detectors. This method was found to be especially suitable for spectroscopic investigations of opaque materials [7, 9], but is certainly not a sensitive method for detecting small concentrations of absorber. The reason for this lies in the acoustic mismatch between the solution and the gas phase. This acoustic mismatch is made clear by the acoustic transmission coefficient α_T, which describes the ratio of the incoming amplitude to the transmitted amplitude of the sound wave. The value of α_T is approximately 10^{-5} for the gas-liquid phase transition. This means that the sound wave produced in the solution is not detected directly by the microphone. The acoustic signal detected by the microphone is not generated until the heat produced in the solution by absorption of optical radiation reaches the liquid-gas phase boundary and produces sound waves in the gas volume by periodic heating. Because of the poor high-frequency characteristics of the microphones, only low frequency modulated light can be used to generate photoacoustic signals with this detection technique. Since the modulation frequencies often lie in the frequency range of perturbing background oscillations, this represents another disadvantage of the "gas microphone technique".

To avoid the acoustic mismatch, directly coupled pressure-voltage converters (piezoelectric crystals) were then used, starting in 1975, to detected the acoustic waves from solutions [6, 7]. The acoustic transmission coefficient for the liquid-solid transition, as is relevant for this method, is approximately 0.4. Because of the very good high-frequency characteristics of piezoelectric crystals, pulsed laser systems can be used to generate the photoacoustic signals [62]. In this way, the background oscillations, which usually lie in lower-frequency regions, can be effectively discriminated from the actual photoacoustic signal by means of electronic filters and time gated measurement of the photoacoustic signals. Another advantage of pulsed photoacoustic spectroscopy lies in the light source itself: because of the easily performed dye change, it makes spectroscopic investigations very much simpler than with a CW laser over a large wavelength range. Photoacoustic spectroscopy of solutions [63] and solids [64] with CW laser systems resulted in only limited success. Besides the high sensitivity of pulsed photoacoustic spectroscopy for determining small absorptions in solutions, this method is used for a number of other spectroscopic studies [65] such as,

— two-photon absorption
— Raman processes
— forbidden transitions
— microscopy

— powders, crystals
— surfaces
— low-temperature experiments
— monomolecular films

The diverse possibilities for application of the photoacoustic effect in spectroscopic investigations have created great interest in this method, since suitable light sources and sensitive detector systems were introduced. This is also manifested in the number of publications [2] concerning the photoacoustic effect and photoacoustic spectroscopy that have appeared since the effect was discovered by A. G. Bell.

The high sensitivity of LPAS is demonstrated by the many measurements that have been performed, for example:

Neodymium in heavy water $\quad\quad\quad\quad\quad\quad$ ($\alpha = 2.2 \times 10^{-5}$ cm^{-1}) [66]
Heavy water $\quad\quad\quad\quad\quad\quad\quad\quad\quad\quad$ ($\alpha > 1.0 \times 10^{-4}$ cm^{-1}) [67]
Light water $\quad\quad\quad\quad\quad\quad\quad\quad\quad\quad\quad$ ($\alpha > 1.0 \times 10^{-4}$ cm^{-1}) [67]
Rare earths in aqueous solutions $\quad\quad\quad\quad$ ($\alpha > 1.0 \times 10^{-3}$ cm^{-1}) [68]
Porphyrin in chloroform $\quad\quad\quad\quad\quad\quad$ ($\alpha = 4.7 \times 10^{-4}$ cm^{-1}) [69]
β-carotene in chloroform $\quad\quad\quad\quad\quad\quad$ ($\alpha = 2.2 \times 10^{-5}$ cm^{-1}) [70]
Neodymium in perchloric acid $\quad\quad\quad\quad$ ($\alpha = 4.2 \times 10^{-4}$ cm^{-1}) [71]

where $\alpha = \varepsilon c$ (ε is the molar absorptivity and c the concentration).

The smallest measured absorptions are indicated in the parentheses. Besides these cited papers, many authors have reported on the sensitivity of photoacoustic spectroscopy and the "thermal lensing" method [2, 5, 6, 19, 24, 72–74]. Detection limits are reached that are smaller by more than a factor of 100 than those in classical absorption spectrometry. Hence it was to be expected that when a system sensitivity of $\alpha \cong 10^{-5}\,\mathrm{cm}^{-1}$ was reached the interesting oxidation states of various actinides could be determined in the sub-micromole concentration range [18, 24]. An essential prerequisite for this is that the actinide ions release the absorbed optical energy primarily in a non-radiative manner and thus contribute to the production of the photoacoustic signals.

2.2 Measuring Principle

The creation of the photoacoustic signal in the solution is based on the conversion of absorbed optical radiation into heat by non-radiative relaxation processes. This principle is illustrated diagrammatically in Fig. 1.

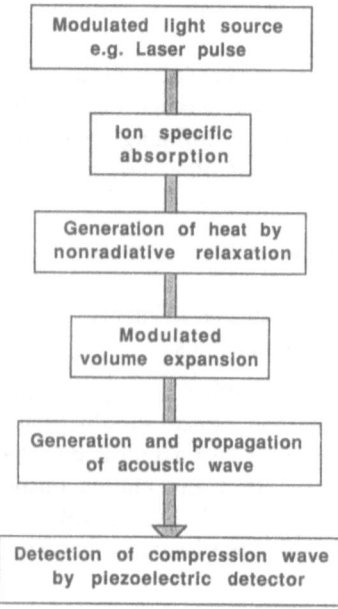

Fig. 1. Generation and detection of photoacoustic signals

The transuranium ions and solvent molecules, located in a higher energy level due to energy-selective photon excitation, release the absorbed energy to the solution in a primarily non-radiative process [19, 66]. As a result, the solution is heated locally in the volume illuminated by the laser pulse and consequently expands. This generates a sound wave with an amplitude proportional to the absorbance α [cm^{-1}]. Starting from the point of origination, the sound wave migrates radially through the solution and is detected by a piezoelectric crystal. Other energy-release processes such as fluorescence and phosphorescence do not contribute directly to the creation of the photoacoustic signal.

The generation of sound waves by the absorption of light was discussed theoretically for the first time in 1963 [75]. Since then, a great deal of theoretical work has been performed [76–83]. Here it is assumed that the generation of sound waves in solutions is caused by electrostriction and the thermoelastic process. In contrast, the radiation pressure itself is negligible in these processes [84].

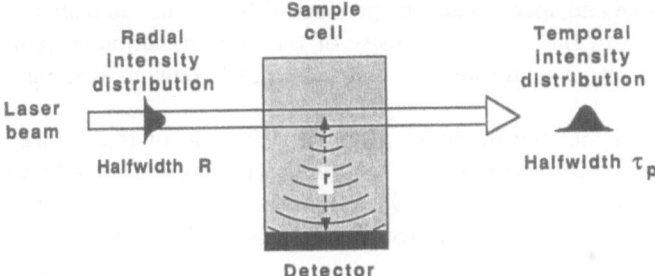

Fig. 2. Geometry of laser beam and photoacoustic signal generation

These photoacoustic experiments are characterized by their geometry, as shown by Fig. 2, and by the ratio of two important time parameters [84–89]. These are the half-width of the laser pulse (τ_p) and the time (τ_a) which the acoustic wave needs to traverse the cylindrical volume illuminated by the laser beam. For τ_a we have:

$$\tau_a = \frac{2R}{v_a},\tag{1}$$

where R is the radius of the laser beam and v_a is the speed of sound in the medium under consideration. For the pulsed laser light source, these important time parameters are as follows:

$$\tau_p = (10\text{–}20)\ \text{ns and}\ \tau_a = \frac{2R}{v_a} = \frac{0.25\ \text{cm}}{1.5 \cdot 10^5\ \text{cm/s}} = 1.7\ \mu\text{s}\ .$$

From this it follows that:

$$\tau_p \ll \tau_a\ .\tag{2}$$

This means that the laser pulse duration is very much smaller than the time that a sound wave needs to emerge from the origination volume. Hence, τ_a represents the characteristic time parameter for the experiment.

The theoretical analysis is performed two-dimensionally, assuming that the acoustic wave is not influenced by the finite length of the sample cell (see Fig. 2). Another simplification results from the fact that the thermal diffusion can be neglected, since

$$\lambda_{(\text{Diff.})} = \sqrt{4\tau_p D}\tag{3}$$

with $\lambda_{(Diff.)}$ = diffusion length [4]. For water, $D \cong 1.4 \times 10^{-3}$ cm^2 s^{-1} [6] and the diffusion length becomes $\lambda_{(Diff.)} \cong 0.1$ μm, which says that

$$\lambda_{(Diff.)} \ll R \tag{4}$$

and, therefore, there are no heat losses due to diffusion during the production of the sound wave. As is further illustrated diagrammatically in Fig. 2, the laser pulses have a Gaussian distribution in time and space. Since the product of two Gaussian distributions is again a Gaussian distribution, the intensity of the laser radiation in time and space can be combined into a single effective distribution, whereby an analytical solution becomes possible.

It can further be assumed that the quantum yield, i.e. the ratio of the number of fluorescence photons and absorbed photons, for the spectroscopically analyzed actinide elements, Am and Pu, e.g. is very low [19], so that the total absorbed energy (E_{abs}) is converted by non-radiative relaxation processes into thermal energy (E_{th}) and thus

$$E_{abs} = E_{th} \tag{5}$$

2.3 Theoretical Description

As mentioned above, from 1963 onwards many theoretical papers have been published concerning the photoacoustic effect and photoacoustic spectroscopy [75–89]. The theoretical models used differ according to the generation of the sound wave as a function of the light source (e.g. high-pressure lamps, CW laser, pulsed laser with microsecond or nanosecond light pulses), according to the medium (e.g. solid, liquid, gaseous phase, optically transparent or opaque material), and according to the detection system. In this section we shall present a theoretical analysis which describes the experiments performed in the authors' laboratory [24]. The discussion describes the pressure produced in a solution by nsecond laser pulses, the propagation of the sound wave in the solution, the transmission process of the sound wave as it crosses over the phase boundaries to the piezoelectric crystal, and finally consider the properties of a detector transforming a pressure variation into a voltage variation.

In a linear approximation, the laser pulse intensity I produces a pressure variation governed by the wave equation [85, 90–91]:

$$\left(\frac{1}{v_a^2} \frac{\partial^2}{\partial t^2} - \nabla^2 \right) p = \left(K_A \frac{\partial}{\partial t} - K_E \frac{\partial^2}{\partial t^2} \right) I \tag{6}$$

where v_a = speed of sound, p = pressure, $I(\vec{r}, t)$ = intensity of the laser pulse (space- and time-dependent), K_A = term for the absorption of heat, K_E = electrostriction term. For K_A we have:

$$K_A = \frac{\alpha \beta}{C_p} \tag{7}$$

with α = optical absorbance, β = volume expansion coefficient, C_p = specific heat at constant pressure, and for K_E we have

$$K_E = \frac{\gamma}{2ncv_a^2} \tag{8}$$

with γ = electrostriction coefficient, n = refraction index, c = speed of light. The linear approximation (Eq. 6) describes a longitudinal oscillatory motion of low amplitude in a compressible liquid. The assumption of low amplitude is justified by the low absorption in the optically thin solutions.

The electrostriction results from the polarization of a dielectricum (here: solution) [93]. The consequence of polarization is a volume contraction which also generates a sound wave and thus makes a contribution to the photoacoustic signal. It must be considered, since the intensity of the laser beam causes a field strength which approaches the magnitude of the intramolecular forces. The field strength produced [94] is calculated from

$$E = \sqrt{\frac{2I}{\varepsilon_r \varepsilon_0 c}} \tag{9}$$

with ε_r = relative dielectric constant and ε_0 = dielectric constant in vacuum, and results in $E \sim 10^6$ V cm^{-1}.

The pressure pulse generated in the solution, therefore, has an electrostriction component and an absorption component:

$$P_A = K_A \frac{\partial \phi}{\partial t}, \tag{10}$$

$$P_E = K_E \frac{\partial^2 \phi}{\partial t^2}, \tag{11}$$

where ϕ represents the solution of the inhomogeneous wave equation [95]

$$\left(\frac{1}{v_a^2} \frac{\partial^2}{\partial t^2} - \nabla^2 \right) \phi = I(\vec{r}, t) \tag{12}$$

For the space and time dependent intensity, we have

$$I(\vec{r}, t) = E_0 f(t) g(\vec{r}) \tag{13}$$

with E_0 = effective pulse energy of the laser radiation, f(t) = time distribution of the intensity with half-width τ_p, g($\vec{r}$) = space distribution of the intensity with half-width R. If the space- and time-distribution is Gaussian, then we have for f(t) and g($\vec{r}$):

$$f(t) = \frac{1}{2\pi} \tau_p \exp(-t^2/2\tau_p^2) \tag{14}$$

and

$$g(\vec{r}) = \frac{1}{2\pi R^2} \exp\left(-r^2/2R^2\right), \tag{15}$$

where $\vec{r}$ is the distance of the pressure wave from the point of origination in cylindrical coordinates. With Eqs. (14) and (15), it follows from Eq. (12) that:

$$\left(\frac{1}{v_a^2}\frac{\partial^2}{\partial t^2} - \nabla^2\right)\phi(t,\vec{r}) = E_0\frac{1}{2\pi}\tau_p\exp\left(-t^2/2\tau_p^2\right)\frac{1}{2\pi R^2}\exp\left(-\vec{r}^2/2R^2\right). \tag{16}$$

Equation 16 represents a partial differential equation which is not solvable directly. If the boundary conditions are known, e.g. large $\vec{r}$, periodicity, origin, observation point, then a Green's function can be constructed which transforms Eq. (16) into a known integral form. The solutions of this integral equation are known as Bessel polynomials. A general solution for the inhomogeneous differential equation Eq. (16) is obtained by introducing the Green's function G:

$$\phi(t,\vec{r}) = \frac{E}{\sqrt{\vec{r}}}\iint G\left(t - t' - \frac{r - \hat{n}\vec{r}'}{v_a}\right)f(t')\,g(\vec{r}')\,d^2\vec{r}'\,dt'. \tag{17}$$

Equation 17 contains the boundary condition for a large $\vec{r}$:

$$|\vec{r} - \vec{r}'|^{-1/2} \cong r^{-1/2} \quad \text{and} \quad |\vec{r} - \vec{r}'| \cong r - \hat{n}\cdot\vec{r}' \quad \text{with} \quad \hat{n} = \frac{\vec{r}}{r'}.$$ Introducing

$$t'' = t' - \frac{\hat{n}\vec{r}'}{v_a} \tag{18}$$

Equation 17 can be simplified to

$$\phi(t,\vec{r}) = \frac{E_0}{\sqrt{\vec{r}}}\iint G\left(t - t'' - \frac{r}{v_a}\right)f\left(t'' + \frac{\hat{n}\vec{r}'}{v_a}\right)g(\vec{r}')\,d^2\vec{r}'\,dt'. \tag{19}$$

With the convolution of the space and time dependent distribution into an effective time profile $\tilde{f}(t)$, it follows from Eq. (19) that:

$$\phi(t,\vec{r}) = \frac{E_0}{\sqrt{\vec{r}}}\int G\left(t - t' - \frac{r}{v_a}\right)\tilde{f}(t')\,dt' \tag{20}$$

with

$$\tilde{f}(t) = \int f\left(t + \frac{\hat{n}\vec{r}'}{v_a}\right)g(\vec{r}')\,d^2\vec{r}'. \tag{21}$$

By insertion it can be verified that the inhomogeneous part is described by the Heavisite function $\theta(t)$ (step function), and for G we have:

$$G(t) = \sqrt{\frac{1}{2\pi}} \frac{v_a}{2t} \theta(t) . \tag{22}$$

Therefore,

$$\phi(t, \vec{r}) = \frac{E_0}{2\pi \sqrt{r}} \sqrt{\frac{v_a}{r}} \int_{-\infty}^{t} \sqrt{t - t'}\, \tilde{f}(t')\, dt' \tag{23}$$

is the solution for an arbitrary effective time-profile. If this effective time-profile is now a Gaussian distribution, since the space and time distributions are Gaussian for the utilized light source, it follows that

$$\tilde{f}(t) = \frac{1}{\sqrt{2\pi}} \tau_e \exp\left(-t^2/2\tau_e^2\right) \tag{24}$$

with $\tau_e^2 = \tau_p^2 + \tau_a^2$. Since $\tau_p \ll \tau_a$, it follows that $\tau_e \cong \tau_a$. Therefore, $\phi(t, \vec{r})$ can be transformed into an analytically solvable function and we have

$$\phi(t, \vec{r}) = \frac{E_0}{2\pi \sqrt{r}} \sqrt{\frac{v_a}{r}} \frac{1}{\sqrt{2\pi}} \tau_a \int_{-\infty}^{t} \sqrt{t - t'} \exp\left(-t'^2/2\tau_a\right) dt'$$

$$\equiv (K/\sqrt{\tau_a})\, \phi_0(t/\tau_a) . \tag{25}$$

The solution of this integral equation is expressed by Bessel polynomials. For ϕ_0 we have:

$$\phi_0(\xi) = \sqrt{\frac{\pi}{8}} |\xi|^{1/2} \left[I_{-1/4}\left(\frac{\xi^2}{4}\right) + \mathrm{sgn}\,(\xi)\, I_{+1/4}\left(\frac{\xi^2}{4}\right) \right] \exp\left(-\frac{\xi^2}{4}\right) \tag{26}$$

with $I_{\pm 1/4}$ = Bessel functions.

$\phi_0(\xi)$ is a function with rising ($\xi \rightarrow -\infty$) and falling ($\xi \rightarrow +\infty$) branches. It describes a compression wave followed by a decompression wave. From Eqs. (25) and (26) and using Eqs. (10) and (11), we find that for the maximum of the pressure wave caused by absorption and electrostriction:

$$P_A(\mathrm{max}) = 0.061 \frac{\alpha\beta}{C_p} E_0 \sqrt{\frac{v_a}{r}} \tau_a^{-3/2} \tag{27}$$

$$P_E(\mathrm{max}) = 0.041 \frac{\gamma}{2ncv_a^2} E_0 \sqrt{\frac{v_a}{r}} \tau_a^{-5/2} . \tag{28}$$

139

The pressure generation described by Eqs. (27) and (28) in solution is finally to be detected by a piezoelectric transducer. Since the detector is not directly introduced into solution in our experiments [18, 24], the sound wave must be conducted by different connecting materials to the piezoelectric detector and thus passes through various phase boundaries. The acoustic transition coefficient represents a measure for the crossing of the acoustic wave at these boundary surfaces [96, 97]. For the crossover of the sound wave from medium 1 into medium 2 we have:

$$\alpha_T = \frac{4 \cdot R_1 \cdot R_2}{(R_1 + R_2)^2} \tag{29}$$

with α_T = acoustic transmission coefficient, R_1, R_2 = acoustic impedances. For the acoustic impedances we have:

$$R_1 = \varrho_1 v_1 \tag{30}$$

$$R_2 = \varrho_2 v_2 \,, \tag{31}$$

where ϱ_1, ϱ_2 = densities of the materials (1, 2), v_1, v_2 = speed of sound in the materials (1, 2). From Eqs. (27) and (28), we then find for the incoming pressure at the piezoelectri crystal:

$$P' = \prod_i \alpha_{Ti}(P_A + P_E) \,. \tag{32}$$

2.4 Piezoelectric Detection

In photoacoustic spectroscopy (PAS) sound waves are detected by microphones or piezoelectric transducers. Especially in photoacoustic experiments in solutions and when pulsed laser systems are used, detection with piezoelectric transducers has some advantages over the microphone technique. These advantages are: simplicity of utilization; possibility of shielding the ambient and background noises by electronic filters; good high-frequency characteristics [98–101] for the use of pulsed dye lasers with their typical short pulse lengths (μs-ns). The main advantage in the use of piezoelectric detectors lies in the low acoustic impendance at the liquid-solid phase transition. The acoustic transition coefficient (Eq. 29) for this phase transition is approximately 0.4 [96]. In contrast, the value is approximately 10^{-5} for the liquid-gas phase transition when a gas-coupled microphone is used. For effective detection of photoacoustic signals from solutions it is important to know the piezoelectric properties of the utilized detector.

Piezoelectric materials have the ability to produce electric charges when an external pressure is applied (piezoelectric effect) or to deform themselves when an electric field is applied (inverse piezoelectric effect) [93, 102]. For small mechanical stresses, that are characteristic of photoacoustic spectroscopy, the piezoelectric effect is linear [103], i.e. the electric voltage generated is proportional to the applied mechanical tension, Z [104], so that:

$$V = K \cdot Z, \tag{33}$$

where K is the sensitivity of the piezoelectric detector. In contrast to natural piezoelectric single crystals such as Rochelle salt, tourmaline or quartz, the material used in PAS are polycrystalline.

Sintered ceramics made of lead-zirconium titanate ($PZT : Pb(Ti_{1-x}Zr_x)O_3$; $x \cong 0.5$) are usually used for photoacoustic experiments [105, 106]. The unit cell of the lead-zirconium titanate has a perovskite structure. Below the Curie temperature (328 °C for the PZT-4 (Vernitron) used by us [24]), the cells are tetragonally deformed, i.e., positive and negative charges are shifted and electric dipole moments are produced. In analogy to ferromagnetism, domains with randomly distributed polarization direction are formed. By the application of an electric field, these can be orientated in a preferred direction, and the sintered polycrystalline ceramic is then remanently polarized. The properties of these anisotropic piezoelectric materials are described by various parameters which depend on the polarization and deformation direction. In the common terminology, the coordinate system shown in Fig. 3 is obtained for the cylindrical piezoelectric crystals [24].

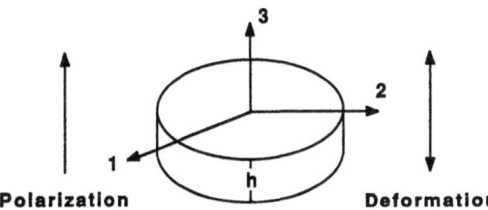

Polarization   Deformation

Fig. 3. Coordination system for cyclindrical piezoelectric detectors

For the detection of acoustic waves it is crucial that the piezoelectric transducer be placed between two mechanical-force-transmitting elements depending on the polarization direction, so that only force components parallel to the polarization direction in the crystal contribute to the charge displacement. In the case of the cylindrical crystals (Fig. 3), these force-transmitting elements have to be applied in the (1,2)-plane, so that the applied mechanical tension and the obtained electrical field strength are in the (3)-direction. This means that only the piezoelectric constants with the (3,3)-coordinate indices are determinative. For an exact detection of sound waves in the PAS it is also important that the transducer be shielded against seismological noises from other materials, ambient noises and light, so as to avoid pyroelectric excitation.

Starting from the plane one-dimensional piezoelectric equations [102],

$$P = Zd + EX \tag{34}$$

$$e = Zs + Ed, \tag{35}$$

where P = polarization, Z = mechanical tension, d = piezoelectric coefficient, E = electric field, X = dielectric susceptibility, e = elastic strain, s = coefficient of elasticity. For the quasistationary case and for the above mentioned geometry (Fig. 3) a charge displacement (q) caused by a pressure change (p) [106] can be obtained by

$$q = K_{33}d_{33}AZ_3 . \tag{36}$$

with

$$Z_3 = \frac{F_3}{A} \equiv p \tag{37}$$

it follows that

$$q = K_{33}d_{33}Ap \tag{38}$$

where K_{33} = coupling factor (dimensionless), d_{33} = piezoelectric charge constant (CN^{-1}), A = surface area (m^2), p = pressure $(N\ m^{-2})$. With the capacitant C of the cylindrical piezoelectric ceramic

$$C = \varepsilon_0 \varepsilon_r \frac{A}{h} \tag{39}$$

and the linkage of charge and voltage via

$$V = \frac{q}{C} \tag{40}$$

we obtain for the voltage:

$$V = \frac{K_{33}d_{33}hp}{\varepsilon_0 \varepsilon_r} \tag{41}$$

with ε_0 = dielectric constant $(A\ s\ V^{-1}\ m^{-1})$, ε_r = relative dielectric constant (dimensionless), h = height of the piezo ceramic (m). As a result, we find the following formula for the proportionality constant designated as the sensitivity in Eq. (33):

$$K = \frac{K_{33}d_{33}h}{\varepsilon_0 \varepsilon_r}. \tag{42}$$

Once the manufacturer's data are known, we can estimate K and thus the voltage obtained at the electrodes of the piezoelectric crystal.

2.5 Calculation of Detection Limit

Now we can calculate the pressure produced in the piezoelectric crystal by absorption and electrostriction, and the resulting voltage. As an example, the calculation shall be performed for the absorption coefficient of H_2O at 503 nm, which is the absorption peak position of Am^{3+} dissolved in purely ionic form. Under the condition of the sensitivity provided by the boxcar measurement technique, the minimum detectable Am^{3+} concentration can then be estimated. The high sensitivity of this measurement technique makes it possible to measure, according to our experience, a signal change that corresponds to a thirtieth to a hundredth of the background signal generated by absorption of H_2O and electrostriction. For the described experiment, the voltage obtained at the detector from the generation of pressure by the absorber and solvent absorption, and also by electrostriction, can be estimated with the following typical data [94, 107, 108]:

$$E \cong 10^{-3} \, J \qquad\qquad r \cong 1 \, cm$$
$$\beta \cong 2.6 \times 10^{-4} \, K^{-1} \qquad\qquad c \cong 3.0 \times 10^{10} \, cm \, sec^{-1}$$
$$\alpha \cong 1.1 \times 10^{-4} \, cm^{-1} \qquad\qquad n \cong 1.33$$
$$C_p \cong 4.2 \, J \, g^{-1} \, K^{-1} \qquad\qquad v_a \cong 1.5 \times 10^5 \, cm \, sec^{-1}$$
$$R \cong 0.125 \, cm \,.$$

From Eqs. (27) and (28), we obtained for the maximum produced pressure change:

$$P_A \,(max) \cong 2.92 \times 10^{-6} \, N \, cm^{-2}$$

and

$$P_E \,(max) \cong 4.39 \times 10^{-7} \, N \, cm^{-2} \,.$$

Under the assumption of an absolute time superposition of P_A (max) and P_E (max) a total pressure change can be estimated:

$$P \,(tot) = P_A \,(max) + P_E \,(max) = 3.36 \times 10^{-6} \, N \, cm^{-2} \,.$$

To detect this pressure change, the sound wave in the utilized arrangement (see Fig. 5) must traverse several phase boundaries, with only a fraction α_T (cf. Eq. 29) of the incident amplitude being transmitted in each case. Based on the literature data [94]:

$$v_a \,(H_2O) \cong 1.5 \times 10^5 \, cm \, sec^{-1} \qquad \varrho \,(H_2O) = 1.0 \, g \, cm^{-3}$$
$$v_a \,(quartz) \cong 5.0 \times 10^5 \, cm \, sec^{-1} \qquad \varrho \,(quartz) \cong 2.2 \, g \, cm^{-3}$$
$$v_a \,(glycerin) \cong 1.9 \times 10^5 \, cm \, sec^{-1} \qquad \varrho \,(glycerin) \cong 1.3 \, g \, cm^{-3}$$
$$v_a \,(PZT) \cong 4.6 \times 10^5 \, cm \, sec^{-1} \qquad \varrho \,(PZT) \cong 7.5 \, g \, cm^{-3}$$

the transmission coefficients can be calculated. We obtain for the different phase boundaries:

$$\text{solution — quartz:} \qquad\qquad \alpha_{T1} = 0.42$$

quartz — glycerin:	$\alpha_{T2} = 0.60$
glycerin — quartz:	$\alpha_{T3} = 0.60$
quartz — piezo ceramic:	$\alpha_{T4} = 0.74$.

The total transmission coefficient is then calculated to be

$$\alpha_T(\text{tot.}) = \prod_{i=1}^{4} \alpha_{Ti} = 0.11$$

To estimate the voltage V obtained from the piezoelectric crystal due to the pressure change p(tot) α_T(tot) at the detector, it is still necessary to calculate the sensitivity K.

From the data given in the literature [105]: $K_{33} = 0.70$, $d_{33} = 289 \times 10^{-12} \text{CN}^{-1}$, $h = 0.6$ cm, $\varepsilon_0 = 8.85 \times 10^{-12} \text{ Fm}^{-1}$, $\varepsilon_r = 1300$, it follows from Eq. (42) that

$$K = 1.05 \text{ V cm}^2 \text{ N}^{-1} ,$$

The voltage obtained at the piezoelectric crystal is then calculated to be

$$V = K\alpha_T(\text{tot}) \, p(\text{tot}) = 3.9 \times 10^{-7} \text{ Volt} .$$

The minimum voltage change ΔV measurable on this "background signal" with the boxcar measurement technique can be:

$$\Delta V = \frac{1}{30} V \sim \frac{1}{100} V . \tag{43}$$

Thus, in the most favorable case a voltage change of

$$\Delta V = 4 \times 10^{-9} \text{ Volt}$$

could be detected. Based on this value we can then estimate the sensitivity of the LPAS system of the author's laboratory i.e., the minimum detectable absorbance α [cm^{-1}], and the detection limit for the Am^{3+} ion in aqueous solution. It follows from Eq. (27) that:

$$\alpha \,(\text{min}) = \frac{\Delta V C_p \tau_a^{3/2}}{0.061 \cdot \alpha_T(\text{tot.}) \cdot K \cdot \beta \cdot E_0 (v_a/r)^{1/2}} , \tag{44}$$

and

$$\alpha \,(\text{min}) \cong 1.0 \times 10^{-6} \text{ cm}^{-1} .$$

With $\varepsilon(\text{Am}^{3+}) = 410$ L mol^{-1} cm^{-1} [19], the detection limit of the Am^{3+} concentration is calculated to be

$$C_{\min}(\text{Am}^{3+}) \cong 2.4 \times 10^{-9} \text{ mol L}^{-1} .$$

This detection limit for the Am^{3+} ion in aqueous solution has to be compared to the experimental speciation sensitivity (see Sect. 5.1). The experimental speciation sensitivity for Am(III) in EDTA is 5.0×10^{-9} mol L^{-1} which corresponds to α (min) $= 3 \times 10^{-6}$ mol L^{-1}, taken the molar absorptivity of Am-EDTA ($\varepsilon = 601$ L mol^{-1} cm^{-1}) into account. This value is three times larger than the theoretical estimated detection limit. The above estimation (Eq. 44) indicates that the detection limit depends directly on the background compensation, which can be performed by a parallel measurement of sample and reference solution.

The electrostriction component of the photoacoustic signal can be neglected for higher absorptions ($\alpha > 5 \times 10^{-3}$ cm^{-1}), since the contribution P_E to the photo-acoustic signal is constant. The background signal is increased by a multiple of the signal resulting from solutions having a high concentration of impurity ions and colloids. This is especially true in natural groundwaters with their diverse consti-tuents [30].

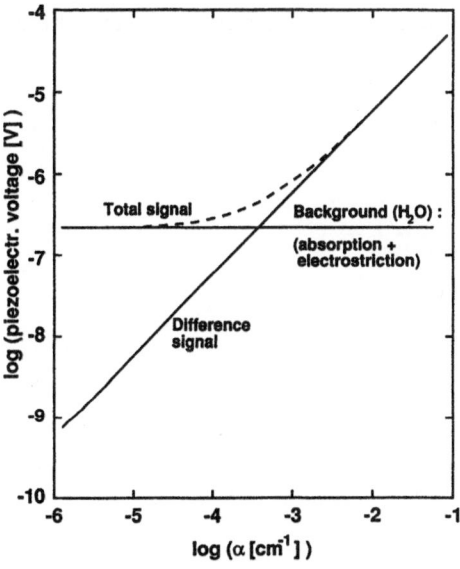

Fig 4. Photoacoustic signals and background compensation as a function of the absorbance (cm^{-1})

Fig. 4 illustrates the dependence of the photoacoustic signals obtained at the photoelectric detector on α for an absolute measurement ($\alpha_a + \alpha_b$) and for a difference measurement (α_a). The background signal is considered as a combined contribution of water absorption at 503 nm and electrostriction. The linear dependence of the photoacoustic signals on the absorption α_a of a given absorber is found if the background signals are differentiated and they are approximated to the background signals in the absolute measurement. Thus, when a difference method is used, the photoacoustic signal is directly proportional to the concentration of a particular absorber (ε = constant; $\alpha \sim c$).

3 Instrumentation of Spectroscopy

3.1 Apparatus

A typical experimental set-up is illustrated in Fig. 5 [24]. Besides the suitable selection of light source, photoacoustic cell and electronics, the experimental set-up includes control and data-processing units so that sensitive experiments can be performed on solutions automatically. A pulsed laser system is used as the light source. An excimer laser pumps a tunable dye laser. The most important excimers are rare-gas/halogen exciplexes [109] such as ArF and XeCl. The emission lines of the excimer lasers lie in the range of 193—351 nm. The excimer laser (EMG 201E, Lambda Physik, Göttingen) is driven with XeCl as a laser medium. With a laser emission at 308 nm and a nearly Gaussian temporal intensity distribution with a half width of about 15 ns, a pulse power of about 350 mJ is obtained. Other laser media are not suitable for the experiments performed here, because the typical pulse shapes do not have a temporal Gaussian distribution but rather a double-pulse characteristic [110]. This would contribute to temporal interference in the photo-acoustic signal formation.

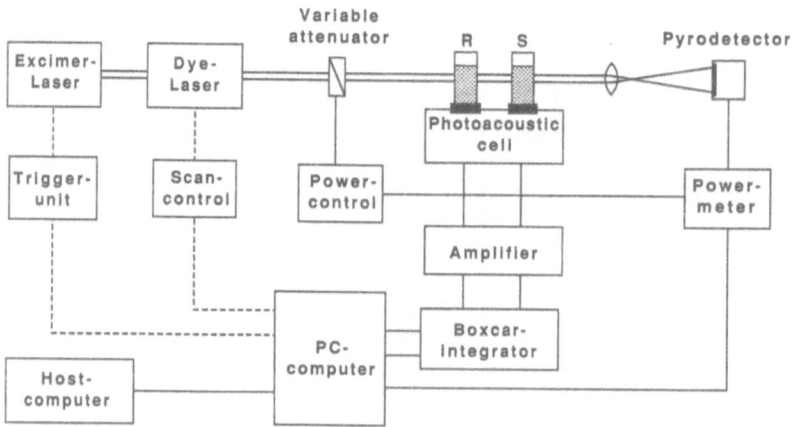

Fig. 5. Schematic layout of laser-induced photoacoustic spectroscopy [24]

The dye laser (FL 2002, Lambda Physik, Göttingen) with a bandwidth of 0.0075 nm at 580 nm (using a grating with 600 lines/mm) supplies pulse powers of up to 60 mJ [111] depending on the dyes used. In the author's laboratory, Lambdachrome dyes (Lambda-Physik, Göttingen) and the recommended solvents are used for nearly all interesting wavelength regions.

For the recording of a spectrum it is, however, a big advantage if a dye change can be dispensed with. The position of the emission spectra of the utilized laser dyes depends on the polarity of the solvent. Thus the changing of the solvent (using dioxan instead of methanol) results in a shift of the emission region of

Coumarin 307 toward shorter wavelengths [112]. The shift toward higher energies is caused by the effect of $n - \pi^*$ transitions (negative solvatochromism [113]). With a suitable concentration of the dye ($1.7 \, \text{g} \, \text{L}^{-1}$ for the oscillator), a shift of about 20 nm was achievable without any power loss. Thus, the spectral range of importance for the spectroscopy of Pu(IV) (460—500 nm) can be tuned without a dye change with the pulse energy required for the experiment.

On the basis of the theoretical description (Sect. 2.3) it is to be expected that an increase of the effective pulse energy of the laser radiation (E_0 in Eq. 27) causes an increase of the photoacoustic signal. However, experiment shows that the increase of the pulse energy is limited, because of the resulting perturbation effects, such as photochemical reactions, multi-photon transitions and localized heating and gas formation. The near ultraviolet region proved to be especially critical with respect to such perturbation effects because of the highly energetic photons. For the investigated substances, an optimal measurement condition is found adequate at pulse energies of up to 3 mJ. These pulse energies are attainable by inserting neutral filters to attenuate the laser beam. In addition, a telescope is inserted for beam widening in the near ultraviolet region. In this way, the local photon density is lowered without decreasing the absolute pulse energy, and the occurrence of the above mentioned perturbation effects is thus reduced.

Since the dye laser power varies as a function of wavelength, following its emission profile [114], a variable attenuator is used for the measurements of wavelength dependent spectra in order to keep the effective pulse energy of the laser radiation constant. With the attenuator (Spindler + Hoyer, Göttingen) the transmission can be varied continuously between 1% and 80%. By measuring the effective pulse energy, the attenuator is driven automatically via an electronic control unit to keep constant a selected energy value during the measurement.

In the given arrangement [24], the power-limited and adjusted dye laser beam falls through the quartz cuvettes with reference and sample solutions positioned on the photoacoustic cell. The two photoacoustic signals are amplified in parallel and electronically filtered with respect to low-frequency interference (AM 502, Tektronix). The amplified and filtered signals are then feed into two 165-segment boxcar slide-in modules and are subtracted from one another by the 162-segment boxcar base unit (EG + G, Princeton Applied Research).

The piezoelectric signal corresponding to the photoacoustic compression wave has a time delay from the laser pulse due to the propagation speed of the acoustic wave through the solution. By positioning the boxcar gate (500 ns) to the first peak maximum the amplitude of the piezoelectric voltage is determined. The dying oscillations of the piezo signal following the first peak are caused by vibrational excitation of the PZT and are ignored for the evaluation of the photoacoustic signal. A storage oscilloscope (468 A, Tektronix) is used to set the boxcar gate and to adjust the sample and reference cuvettes in the laser beam.

The pulse energy of the laser beam is measured by a pyroelectric detector (ED100, Gentec) for normalization of the individual experimental data. The boxcar output signals and energy data are read into a personal computer for further on-line data processing. Recording of the spectra over a wavelength range is performed automatically.

3.2 Piezoelectric Detector and Photoacoustic Cell

Since 1975 [63, 115], PZT transducers have often been used for the detection of photoacoustic signals from solutions. Many different piezoelectric detectors have been developed for these experiments [67, 70, 116–121]. Piezoelectric ceramics in tubular form are frequently used. Then the detectors serve simultaneouly as a container for the sample solution. Despite many proposals for the construction of a piezoelectric detector, it is necessary to develop a special cell and detector unit to perform photoacoustic experiments on highly radioactive corrosive solutions. The following requirements had to be taken into special consideration in the design:

— The smallest possible number of phase transitions for the acoustic wave from the solution to the PZT.
— Only one maximum force component parallel to the polarization direction of the crystal.
— A good shield for PZT against ambient noise and electrical "pick-up". This is especially important when pulsed laser systems are used, where strong electromagnetic pulses are produced during the discharge.
— An effective shield for PZT against scattered light. Through the pyroelectric effect and also through absorption in the PZT itself, scattered light can produce perturbation signals which make significant contributions to the actual photoacoustic signals that are to be detected from the solution. The bare piezoceramic material is nearly optically impermeable, so that absorption of scattered light causes an especially serious problem.
— No chemical interactions between the sample solution and PZT. Especially aggressive, acidic solutions can dissolve constituents out of the PZT material. The impurities can cause photoacoustic perturbation signals which create much interference in weakly absorbing solutions. The use of bare piezoceramic disks or piezoceramic tubes as a sample container represents an undesirable source of contamination. It is found that even surface-treated (gold-plated) piezo tubes are not resistant to many acids.
— No contact of PZT with the radioactive solutions. In some circumstances, contaminated detectors can impair the experimental result by further contaminating other samples.
— The resulting photoacoustic signals with their small amplitude (10^{-6} V — 10^{-9} V) should be adequately preamplified so that they can be measured with the boxcar, and perturbing components from other frequency ranges should be filtered out effectively by electronic means.
— The detection system is constructed in such a way that the sample can be changed without difficulty and contamination. This is very important for routine measurement procedures.

The photoacoustic cell shown in Fig. 6 is developed on the basis of the requirements listed above. A difference method is made possible by the arrangement of two identical detector and amplifier units. The photoacoustic cells are accommodated in a black anodized metal housing. The closed metal housing represents a high-frequency-tight Faraday cage. In this way the detectors and the associated electronic

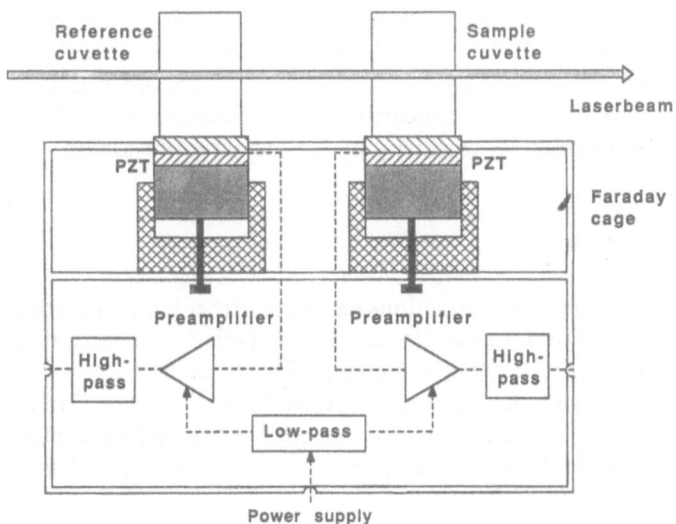

Fig. 6. Photoacoustic single beam dual detector system [18]

components are well shielded against electromagnetic interference radiation. In the housing there are two detector units with two preamplifiers and high-pass filters. Holders for the sample and reference cuvettes are on top of the photoacoustic cell. The preamplifiers are supplied by an external voltage source and high-frequency pulses on the voltage supply lines caused by the excimer laser system are filtered out electronically in several stages and thus are not scattered into the preamplifiers.

The detector unit, as shown in Fig. 6, consists of the sample and reference cuvettes which are acoustically coupled with glycerin [122] to each element consisting of quartz plate, piezoceramic disk and metal cylinder. The sample cuvette itself and the quartz plate on the top as well as the metal cylinder on the bottom represent the components transferring force parallel to the polarization in the (3)-direction (see Sect. 2.4). With the quartz plate on the top, the piezoelectric crystal can be enclosed in the metal housing which provides effective shielding against electrical "pick-up". A quartz plate was chosen because the best acoustic transmission is ensured by the coupling of the quartz cuvette with glycerin. The quartz plate is coated with an Al film to shield the detector against scattered light from the solution or from the walls of the sample cuvette. This film has a reflectance of about 90% over the entire visible wavelength region. For the electrodes of the PZT a thin copper foil is used, which is glued to the piezoceramic disk with conducting adhesive.

The detector unit constructed in this manner offers optimal conditions for photoacoustic experiments on actinide solutions. Since the photoacoustic signals are very small (10^{-6}–10^{-9} V, see Sect. 2.5), they have to be amplified for the boxcar measurement. At the same time, it must again be made certain that electrical "pick-up" is avoided. Therefore, the signals have to be amplified directly in the photoacoustic cell. A broad band preamplifier with extremely low electronic self-noise is developed for that purpose. After constant primary amplification (by about

1000 times), a second variable amplification (by up to 10 times), depending on the amplitude of the photoacoustic signal, is possible. Low-frequency interference signals caused by ambient noise are filtered out by a high-pass filter. These low-frequency interference signals would otherwise make noncorrelated contributions to the photo-acoustic signal and thereby worsen the sensitivity of the measurement system.

3.3 Variation of Instrumentation

The instrumentation shown in the previous sections has been further improved in our laboratory for a better compensation of background signals [40]. At the same time, based on the concept of the instrumentation shown in Fig. 5 the new sets of the spectrometer have been developed in Argonne National Laboratory [20, 21, 25] and Harwell Laboratory [22, 26, 27] and Lawrence Livermore Naional Laboratory [23] for the radiochemical speciation of aqueous actinide ions. These activities are summarized as follows.

3.3.1 Dual Beam Single Detector System

For a better compensation of background signals of water, a dual beam single detector is introduced [40] so as to avoid the effect of different characteristics of two piezoelectric detectors used for sample and reference. A schematic diagram of the dual beam single detector system introduced into LPAS is shown in Fig. 7. The instrumentation is the same as that given in Fig. 3. Sample and reference cuvettes are put on the same detector and the laser pulse is directed to the two cuvettes alternately. The laser beam switching is achieved by two mirrors in a staggered arrangement, one of them is laterally displaced by an eccentric motor drive and thus the moving mirror reflects the laser beam to the second cuvette. The laser is triggered at the dead point of the eccentric drive and hence the mirror is at rest. The spatial adjustment of laser beams at each cuvette is better than 0.1 mm. The piezoelectric signals are amplified, fed alternately into two boxcar channels and processed in the usual manner as described above (Sect. 3.2). In the experiment

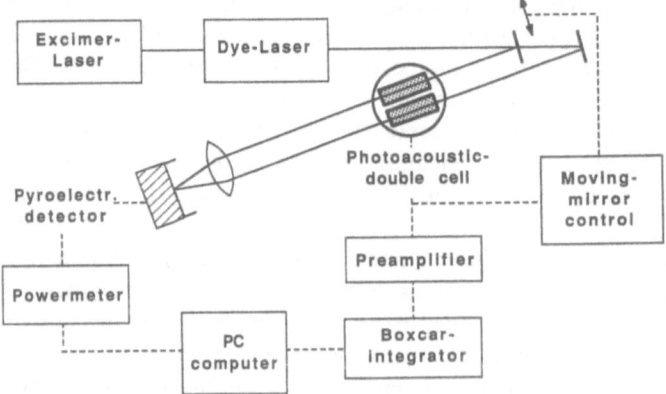

Fig. 7. Schematic layout of laser-induced photoacoustic spectroscopy with dual beam single detector system [40]

with the system using pure water in both cuvettes, piezoelectric signals detected in a wide spectral range are found to be almost identical in amplitude and time dependence. This is not always the case when two separate detectors are used with individually varying electronic properties.

3.3.2 Instrumentation for High Temperature Experiments

A block diagram of the experimental apparatus is shown in Fig. 8 [20, 21, 25]. The excitation source is an excimer pumped dye laser [Lambda Physik EMG 201 MSC excimer (XeCl); FL3002 dye laser]. This apparatus is capable of producing energies in excess of 40 mJ at 480 nm with a resolution of <0.2 cm^{-1} (<0.01 nm at 480 nm), has a pulse width of 28 ns, and can be operated at repetition rates up to 80 Hz. The overall tuning range of the dye laser is ~ 300 to 1000 nm, although it requires approximately 20 dyes to cover this region adequately. Coumarin 480 dye (460–490 nm range) is employed for the LPAS studies of Ho^{3+} and Pu^{4+}, while Coumarin 500 dye (490—550 nm range) is used for Am^{3+}.

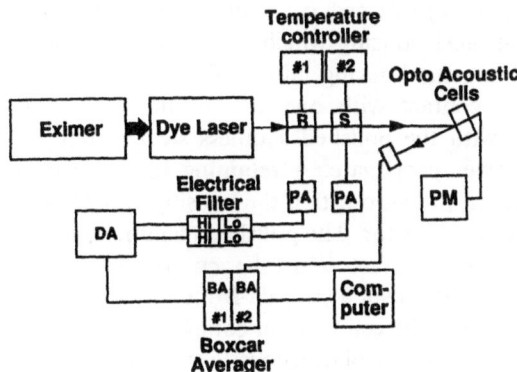

Fig. 8. Schematic layout of high temperature laser-induced photoacoustic spectroscopy (R & S: reference and sample cells, PM: power meter, PA: preamplifier, and DA: differential amplifier) [20]

Before entering the sample and reference cells, the output of the dye laser is spatially filtered by an iris that confines the beam to a cross-sectional area of approximately 2 mm^2. This procedure also reduces the amount of non-tunable amplified spontaneous emission (ASE) concentric to the laser beam. After emerging from the apparatus, the laser light is then monitored in two ways: (1) power meter (Scientech calorimeter 36-0201, NBS traceable) from which a 5% reflection is incident upon a (2) pyroelectric energy detector (Molectron J3-05) used for normalization of raw LPAS data.

A cross-sectional view of one cell assembly of the differential photoacoustic apparatus is shown in Fig. 9. Optical access to the sample solution contained in a 1-cm cuvette (an all-quartz modification of a cuvette from Hellma Cell Inc. No. 221, which reduces the probability of cell fracture) is provided by 6.4 mm apertures in a thermostated aluminium housing. Temperature control is provided by four cartridge heaters (Hot Watt, 100 W/120 V) mounted at the four corners of the aluminium block and connected in series to a 10 A/120 V triac (LOVE Controls Corp). The power output of the triac is controlled by a proportional, integral,

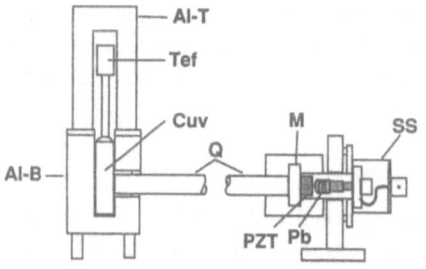

Fig. 9. High temperature photoacoustic detector system (Al-T & Al-B: heated Al block, Cuv: cuvette, Tef.: teflon swagelok, Q: quartz rod, M: front surface mirror, SS: stainless steel housing and Pb: lead disc)

derivative temperature controller (LOVE Controls Corp. Model 300) in combination with J-type thermocouple inputs (system stability = ± 1 °C). A temperature-limiting switch is used to prevent accidental over-heating. The cuvette stem, sealed by a Teflon fitting, is heated to a slightly higher temperature than the bottom portion of the cuvette by yet another thermostated aluminium block. This differential heating procedure prevents refluxing of the solution.

Signals generated following the absorption process are detected perpendicular to the excitation by a piezoelectric transducer (PZT) (Transducer Products LTZ-2: 9 mm dia, 5 mm thick) mounted in a stainless steel housing which is electrically isolated from the aluminium heating block. Within the housing, ultrasonic fluctuations are dampened by a lead disc in electrical contact with one side of the PZT. The remaining PZT electrode is in contact with the polished stainless steel membrane (~ 1 mm) through which the sound waves propagate. A retaining spring is then used to keep all components under constant pressure within the housing and provide electrical continuity to the BNC electronic connector. The polished membrane of the stainless steel housing is held in acoustic contact with the cuvette by means of a quartz rod (1 cm dia, 10 cm or 61 cm length) via a silvered-front surface mirror. This arrangement allows acoustic contact with the cell without complications arising from temperature influences on the transducer and photoacoustic signals generated by scattered light. Acoustic contact is improved by use of a thin layer of grease (Halocarbon 25-S) at each interface leading to the transducer. This grease is chosen on the basis of its high temperature stability, its similar refractive index with quartz, and its minimal optical absorbance.

The outputs of each PZT (reference and sample signals) are preamplified by use of operational amplifiers (Analog Devices 50 K) with variable gain (noise $\sim 5 \, nV/Hz^{1/2}$). Variable gain allows the response of each photoacoustic cell (containing identical solutions) to be matched prior to analysis. The preamplified signals are high- and low-pass filtered to eliminate extraneous environmental noise sources and then differentially amplified (Tektronix AM502, CMRR — 50 dB at DC to 50 kHz). The differential output (analyte = sample — reference) is fed into a boxcar integrator [Stanford Research Systems (SRS) model SR250 Boxcar Averager], the gate of which is centered on one of the first maxima of the PZT response (25–35 µs for 10-cm rod, 130–140 µs for 61 cm rod). The trigger for the boxcar is provided by a photodiode that monitores the excimer repetition rate. Averaging of the signal is performed over a few to several thousand laser pulses, depending on the desired signal-to-noise ratio. The dye laser energy (LPAS signals are linear at laser energies of 3–4 mJ/pulse for Am^{3+}, up to 25 mJ/pulse for Ho^{3+}) is simultaneously recorded using a second boxcar. Averaged outputs from each boxcar are then digitized (SRS model SR 245 computer interface) and recorded on a computer (IBM PC/XT, ANL modified SRS model SR265 data acquisition software) as a function of excitation wavelength. The PZT response is divided by the laser energy to obtain the normalized LPAS spectrum for the absorbing species.

3.3.3 Dual Beam Two Detector System

A LPAS facility built at Harwell [22, 26, 27] is shown diagramatically in Fig. 10. A Lambda Physik, EMG 101 MSC XeCl excimer laser pumps a Lambda Physik, FL2002 wavelength tunable dye laser (available through Coherent UL Ltd.). The

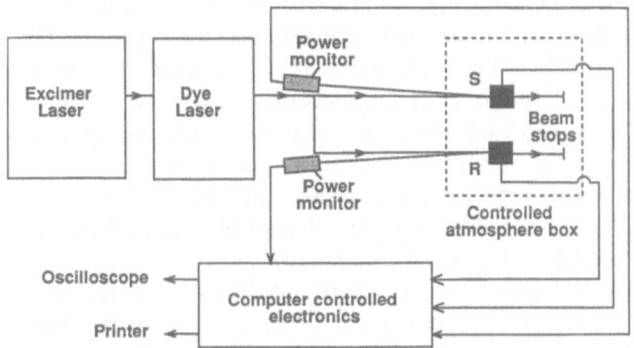

Fig. 10. Schematic layout of laser-induced photoacoustic spectroscopy with dual beam detector system (R & S: reference and sample cells) [22]

emerging beam is split, using an appropriate 50% beam splitter (Technical Optics Ltd, Onchan, Isle of Man) in conjunction with a mirror, to produce a dual beam analytical system. Measurement of absorption spectra, over a wavelength range of 312 nm to 985 nm, is complicated by the need to change beam splitters and laser dyes. Six beam splitters are required, as the degree of S and P polarisation of the dye laser beam varies with wavelength. Each beam splitter functions quite acceptably up to ± 20 nm or more each side of its specified range. Seven dye flow cells each loaded with a different dye are available to minimize time losses whilst changing dyes. The lifetime of the excimer laser gas fill has been considerably extended, by the adaption of the laser gas fill system to allow the operator to bleed in some extra HCl/He when the power of the excimer starts to drop. An additional valve is also installed in the excimer laser head to allow the laser head tank to remain isolated whilst the rest of the system is evacuated.

The dye laser pulse energy in both beam lines is constantly monitored using pyroelectric joulemeters (Type J.3, Molectron Detector Inc.). The photoacoustic signals from both the sample and reference cells (10 mm or 40 mm path length stoppered UV silica cells, Pye Unicam) are amplified and fed with the joulemeter outputs into the data acquisition system. This allows full spectral and pulse to pulse correction of both laser power variation and solvent absorption to be made. While there are very small variations in the energy response of the joulemeters over the wavelength range 312 nm up to 985 nm the two joulemeters are interchangeable. Automatic compensation for this variability occurs because of the dual beam operation. The joulemeters are positioned at as small an angle to the beam lines as possible to reduce the possibility of introducing beam energy variations due to S and P polarisation differences.

The electronics for controlling the lasers, dye laser wavelength scanning and data acquisition are comprised of two parts: the laser control system and the Computer. In the laser control system the signals to trigger the excimer laser and to scan the dye laser are converted to light signals to allow the electromagnetically noisy excimer laser to be electrically isolated. Up to 8 channels of data can be simultaneously collected and amplified. Any two of them can also be monitored by a digital storage oscilloscope (type 468, Tektronix). One data channel is assigned to monitor the excimer laser power, another two the power of the dye laser beams and a further two the piezoelectric detector signals from the sample and reference cells. The remainder are spare channels. The eight channels of information are fed through a 12 bit ADC into a Z8 single chip microprocessor. Its fast operation allows rapid input/output operations and some data manipulation. The data are then transferred to the Apricot P.C. (Applied Computer Techniques) for display and storage. Currently background corrected spectra with a maximum of 4096 laser pulses/point with the laser firing at 50 Hz for a wavelength scan of 50 nm with 0.1 nm resolution, or any equivalent combination, can be acquired.

The photoacoustic sample cell/detector assemblies are inside a controlled atmosphere glove box. This glove box has two functions, firstly to allow experiments to be carried out in the anaerobic conditions expected in deep repositories, and secondly to act as containment for the radioactive materials. The nitrogen atmosphere glove box has been fitted with sufficient windows to allow ready adaptation of the LPAS system to carry out other photothermally-based spectroscopy techniques, namely thermal lensing spectroscopy and photothermal deflection spectroscopy.

Initially the LPAS system is operated in a single beam mode to investigate photoacoustic signal characteristics and to optimize photoacoustic signal amplification and gating. A special low noise, high gain preamplifier is developed. This is comprised of a low noise FET input amplifier (2N6550 Teledyne Crystalonics) followed by two further gain stages. The peak voltage gain at approx. 500 kHz is about 90 dB. A further 40 dB of switchable voltage gain is also available for the photoacoustic signals prior to data acquisition. This particular unit also removes the negative going components of the photoacoustic signals so that the analogue to digital conversion stage which follows can be operated with the widest dynamic range.

Dual beam operation is then investigated. The main problem is to produce two matched photoacoustic detection systems, each comprising a cell, detector and amplifier. The dual beam cell arrangement is shown in Fig. 10, and the photoacoustic cell detector arrangement developed is shown in Fig. 11. Each cell, a stoppered quartz cuvette, is coupled to the diecast box using optical coupling jelly. The piezoelectric transducer is mechanically held against the lid of the diecast box, which acts as the signal ground. The signal output is taken off the transducer rear face, which is in contact with a quartz flat (microscope slide) that acts as an insulator for the metal clamp. The clamp holds the component stack together. The configuration described maximized the sensitivity and screened the detector/

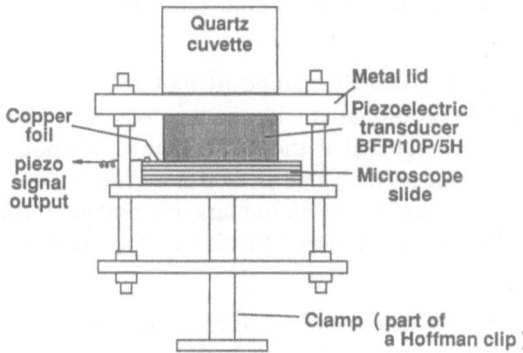

Fig. 11. Photoacoustic detector system [22]

preamplifier assembly from electrical noise, particularly the firing of the excimer laser. A range of lead zirconate titanate piezoelectric devices is investigated. A 500 kHz transducer (Type BFP/10P 5A Vernitron) is found to be the most suitable.

4 Characterization of Spectroscopic Operation

The spectral work performed with the instrumentation given in Fig. 5 is characterized and the results are discussed here. The discussion is also relevant to the other LPAS setups given in Sect. 3.2.3 and 3.3.3.

4.1 Data Processing

The objectives of the data processing can be divided into three areas:

— Averaging of photoacoustic signals to eliminate random fluctuations, which are caused primarily by the fluctuation of the pulse energy of the laser radiation, by electronic noise and by pick-up of non-filterable low-frequency interference signals.
— Detection and elimination of noncorrelatable photoacoustic signals which are produced, for example, by dielectrical breakdown and gas formation effects [6].
— Normalization of the measured data with the pulse energy of the laser radiation.

The random fluctuations are averaged out by reading at least 100 boxcar data for a given wavelength and averaging the boxcar data in the computer. At the same time, the pulse energy of the laser beam, measured with the energy measuring device, is read into the computer and averaged over the appropriate number of data items. However, the program sequence, i.e. the data processing, is so designed that the signals that are caused by interference effects and whose amplitude cannot be correlated with the "true" photoacoustic signal are recognized by the computer and eliminated. The recognition and elimination of outlier signals are accomplished in two steps. A prerequisite for this is that the very first value of a measured point be obtained without outliers:

— The first step prevents the reading in of primarily significant outliers. For this purpose, each new boxcar signal read-in is compared with the previous one. The limits are so chosen that an outlier is recognized if there is a change of about 10% in the two consecutive boxcar signals. This value is then not read in.
— The second step provides for the elimination of read-in outliers, i.e. outliers that are not recognized in the first step. This is accomplished by means of a statistical outlier test.

For the statistical test, the program determines the arithmetic mean $\bar{x}$ of the $n = 100$ read-in boxcar signals and also the standard deviations. Then the r (99.9) test [123, 124] for outliers is performed and they are eliminated. The process is performed until self-consistency is attained or until a minimum of 49 boxcar signals, since the test is usable only if there are at least 50 data items. The averaged outlier-free value for a constant wavelength j is finally stored with the associated average pulse energy in the working memory. At the beginning of the reading in of the boxcar signals for the next wavelength status, the stored mean value is compared with the newly read-in one. If the change is $\geq 30\%$, this measurement value is recognized as an outlier and is not read in. The variance of this change is chosen to be relatively large, since a signal change due to an increase or decrease of the absorption coefficient in the region of an absorption band is also possible. At the end of the measurement, the wavelength-dependent data are stored in the form

wavelength	mean boxcar signal	mean pulse energy	photoacoustic signal
λ_j	$\bar{X}_j$	L_j	$\bar{X}_j/L_j$

After scanning the chosen spectral range the data are stored on disk and the photoacoustic spectrum is plotted.

The spectrum scan can be repeated automatically several times. This multiscanning allows further statistical improvement of the baseline scattering, which is especially important in colloidal solution with noisy background.

4.2 Calibration of Signal Magnitude

As a first investigation, the magnitude of the photoacoustic signals of the Am^{3+} ion is estimated. In order to keep the magnitude of a background signal α_b due to water absorption and electrostriction small compared to the absorber signals (see Sect. 2.6 and 2.8), the concentration of the ions to be investigated is chosen so that

$$\alpha \geq 6 \times 10^{-3}\ cm^{-1}\ .$$

For the Am^{3+} solution utilized for this purpose, we obtain a ^{241}Am concentration of

$$C(Am^{3+}) = \frac{\alpha_a}{\varepsilon} = \frac{6 \times 10^{-3}}{410}\ mol\ L^{-1} = 1.5 \times 10^{-5}\ mol\ L^{-1}\ .$$

0.1 M $HClO_4$ is used as a solvent, since this represents a non-complexing medium for Am^{3+}, in which Am^{3+} is present in purely ionic form [125] and the molar absorption coefficient is a well defined quantity: $410 \ L \ mol^{-1} \ cm^{-1}$ [19]. In the region of the main absorption band of Am^{3+} at 503 nm, the photoacoustic spectrum shown in Fig. 12 is recorded. The obtained absolute boxcar signal (V) is plotted versus the wavelength (nm). A dye-laser pulse energy of about 2 mJ is used in this measurement. The dye-laser beam is positioned so that it falls through the solution about 5 mm above the bottom of the cuvette. This means that the distance from the volume generating the pressure wave to the piezoelectric crystal is approximately $9 \sim 10$ mm.

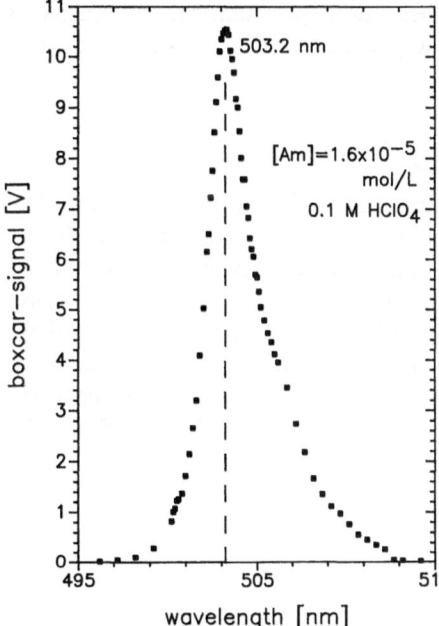

Fig. 12. Photoacoustic spectrum of the Am^{3+} ion $(1.5 \times 10^{-6} \ mol \ L^{-1})$ in 0.1 M $HClO_4$ for the photoacoustic signal calibration [24] (see text)

The spectrum exhibits a boxcar signal of 10.8 V at the peak maximum at 503.2 nm. For comparison, the theoretical value expected for these test conditions is calculated. With Eq. 27 and using the values calculated in Sect. 2.4 for the total transmission coefficient ($\alpha_T = 0.11$) and for the "sensitivity" ($K = 1.05 \ V \ cm^2 \ N^{-1}$) of the transducer, we get the following value for the voltage V_{Piezo} arising at the detector under test conditions:

$$V_{Piezo}(Am^{3+}) \cong 14 \times 10^{-6} \ V .$$

Taking into consideration the applied preamplification (10^4 times) and the amplification at the boxcar (100 times), a theoretical boxcar signal of about 14 V is obtained. This value is close to the boxcar signal of 10.6 V measured experimentally. The deviation of the theoretical value from the experimental value results from uncertainties in the input data, boundary conditions assumed for the theoretical description and inaccuracies in the beam positioning and in the amplifications.

4.3 Correlation of PA Signal to Laser Pulse Energy

According to the theoretical considerations presented in Sect. 2.3, it is expected that the photoacoustic signal, and thus also the boxcar signal, is directly proportional to the pulse energy of the laser beam under constant conditions of the experimental system. Because the pulse energy varies as a function of wavelength on the basis of emission profiles of laser dyes, the individual boxcar values must be normalized to the laser pulse energy. After this normalization, the photoacoustic spectrum should be identical with the corresponding absorption spectrum. Measurements to examine the direct proportionality between the photoacoustic signals and the pulse energy are performed with two solutions, namely:

— 1.3×10^{-6} mol L^{-1} ^{241}Am(III) in 10^{-3} M EDTA
— 1.8×10^{-5} mol L^{-1} ^{238}Pu(IV) in 1 M HClO$_4$.

The Am-EDTA solution is a chemically stable and optically pure solution [19], since EDTA is a strong complexing agent for Am [28]. In contrast, the Pu-HClO$_4$ solution is a strongly colloid-containing solution [126] which is prepared by dissolving solid Pu(OH)$_4$ in 1 M HClO$_4$. This means that this solution has scattering centers for the laser beam and therefore is not optically "pure". Accordingly, with these solutions we can anticipate a broadening of the laser beam and thus an acoustic-wave origin which varies in volume and intensity distribution. The measurements are performed at the peak maximum of the absorption band, i.e., at 506 nm [19] for the Am-EDTA solution and at 476 nm [126] for the Pu-HClO$_4$ solution.

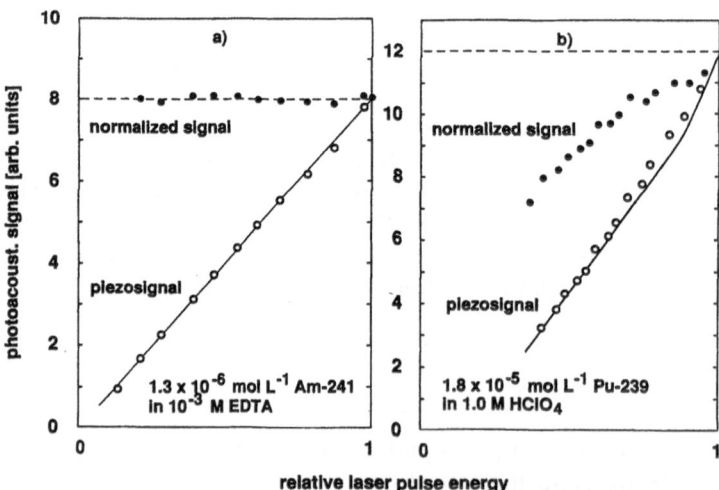

Fig. 13. Correlation of the photoacoustic signal with the normalized laser pulse energy for Am(III) in 10^{-3} M EDTA and Pu(IV) in 1 M HClO$_4$ (containing Pu(IV)-colloids) [24]

The measurement points (O) obtained by varying the pulse energy of the laser radiation are plotted versus the normalization factor $L(max)/L_j$ in Fig. 13. The normalized values (●), also plotted in this figure, would have a constant value and lie on a straight line (———) if there is a directly proportional dependence. It can be seen that the proportionality is satisfied only for the pure $^{241}Am^{3+}$ solution and thus the normalization performed in this manner is justified. This relationship does not hold for the colloid-containing $^{238}Pu(IV)$ solution [126] due to scattered-light effects which are dependent on the concentration and size distribution of the scattering centers [127].

In practice the pulse energy is kept constant within the measuring region. The measurement of the pulse energy by means of a pyroelectric detector enables energy-dependent positioning of a variable attenuator in the dye laser beam with a specially developed control unit, so that a constant transmission is achieved. Control is exercised with an accuracy of about 5%, which is attributable to the system-related fluctuation of the pulse energy. Within this fluctuation range, the normalization described in Sect. 4.1 is valid.

4.4 Difference Method for Background Compensation

To characterize the difference method used, measurements are carried out with pure (doubly distilled) water. Assuming identical test conditions in experimental cells A and B, when a difference is formed from measured signals from identical absorbers the boxcar signals should fluctuate about the value 0 and should be independent of the absolute magnitude of the signals from the individual experimental cells. The compensation relationship is given by

Boxcar signal (A) — Boxcar signal (B) =
const. $[\alpha(A) \times E(A) \times D(A) \times V(A) - \alpha(B) \times E(B) \times D(B) \times V(B)]$,

where α = absorbance (here: water and wall absorption), E = pulse energy of the laser beam, D = sensitivity of the detector $(K\alpha_T)$, V = electronic amplification. Since the contribution of electrostriction is relatively small (see Sect. 3), the effect after its incomplete compensation between the cell A and B is considered as negligible.

Figure 14 illustrates the results of the measurements for two different wavelength regions under the characteristic test conditions (see Sect. 4.2). These wavelength regions represent a region of very small water absorption ($\alpha \cong 8.1 \times 10^{-5}$ cm^{-1}) at 480 nm and a region of large water absorption ($\alpha \cong 1.3 \times 10^{-3}$ cm^{-1}) in the near IR (830 nm). After carefully balancing the different sensitivities of the detectors by means of appropriate electronic preamplification and after normalization of the laser pulse energy over a measurement range, a nearly constant background fluctuating around the zero point is obtained. The photoacoustic signals for the absolute measurement and for the difference measurement with regulated pulse energy are shown in Fig. 13. The difference measurement leads to wavelength-independent signals only slightly fluctuating from 0, which provides satisfactory conditions for spectroscopic analysis. The photoacoustic signals of the absolute measurement now exhibit a rise

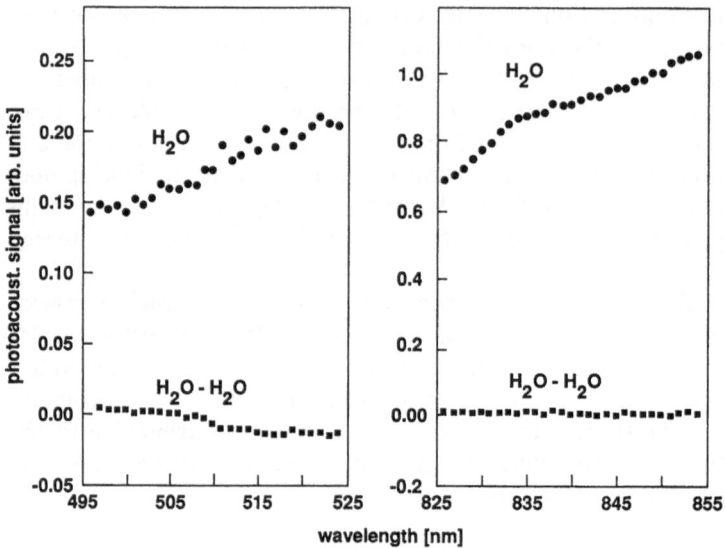

Fig. 14. Photoacoustic spectra of water in VIS and NIR regions: normal absorption of H_2O (upper spectra) and after compensation of H_2O against H_2O (lower spectra) [19]

towards higher wavelengths. Water absorption spectra obtained by other authors [5, 128] show comparable characteristics in these wavelength regions.

4.5 Correlation of PA Signal to Absorber Concentration

To demonstrate the analytical applicability of LPAS in a wide range of absorber concentrations, the linear relationship of the photoacoustic signal to the absorber concentration is measured under constant experimental conditions [19]. Solutions used are Am^{3+} in 10^{-3} M EDTA (Ethylendiamine tetraacetic acid) at pH = 3 and Pr^{3+} in 0.1 M HNO_3. The Am-EDTA complex is a most stable species in aqueous solution with the absorbance of 601 ± 6 L mol^{-1} cm^{-1} at 506.5 nm and therefore is used for the lower concentration range ($< 10^{-6}$ mol L^{-1}). The Pr^{3+} ion with its absorbance of 3.54 ± 0.15 mol^{-1} cm^{-1} at 482 nm is used for the higher concentrations ($> 10^{-4}$ mol L^{-1}). The calibration of photoacoustic signals (μV mJ^{-1}) relative to absorbance is illustrated in Fig. 15, which shows a linear correlation in the absorbance range over three orders of magnitude. Together with the experiment shown in section 4.5, the pulse height of the photoacoustic signal is proved to be lineary correlated with the absorbance (α) as well as the effective laser pulse energy (E_0) under given experimental conditions. This relationship can be simplied from Eq. (44):

$$V(\mu V) = \text{const. } E_0(\lambda) \, \alpha(\lambda) \tag{45}$$

which can be written again for the normalized laser pulse energy:

$$V(\mu V \, mJ^{-1}) = \text{const. } \alpha(\lambda) . \tag{46}$$

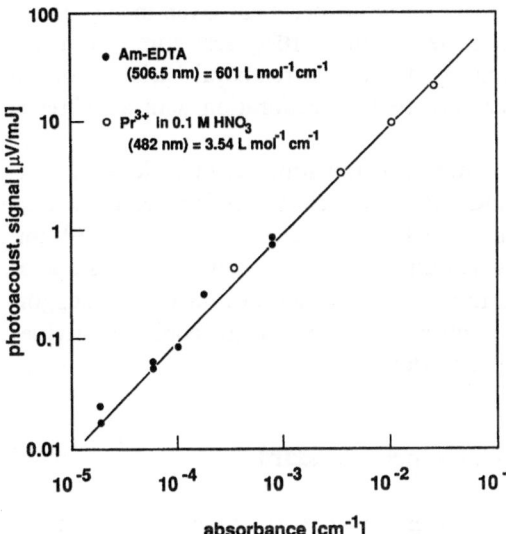

Fig. 15. Linear correlation of the photoacoustic signal with the absorbance of Am(III) in 10^{-3} M EDTA solution for lower and Pr(III) in 0.1 M HNO$_3$ for higher absorbances [19]

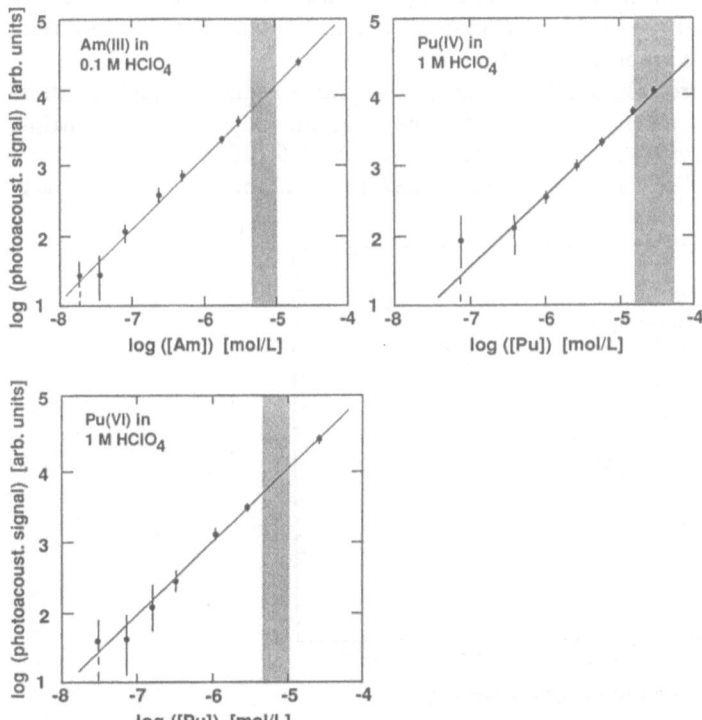

Fig. 16. Calibration of photoacoustic signals as a function of the concentration of Am(III), Pu(IV) and Pu(VI) [18, 24]

The direct verification of Eq. (46) is the calibration experiment given in Fig. 15. In this figure the measured data points have about $\pm 10\%$ scattering from the normalized calibration line. This is primarily due to the temporal variation of the laser pulse energy which can be further corrected by calibration with a reference standard (see below).

The calibration of a similar linear relationship for some actinide ions in non-complexing solution, i.e. Am^{3+}, Pu^{4+} and PuO_2^{2+}, are shown in Fig. 16. These ions are measured at their maximal absorption peaks at 503 nm, 475 nm and 830 nm, respectively [24]. All experimental points remain on the linear curve with a slope of one, suggesting the spectroscopic applicability of LPAS for a wide range of wavelength. The shadowed areas indicate the concentration region of spectroscopic sensitivities attainable by conventional absorption spectroscopy.

4.6 Simultaneous Calibration by a Reference Standard

An improvement introduced in the LPAS operation is a simultaneous calibration of the detection signal to increase the precision of the measurement [40]. The photoacoustic signal of the sample is normalized by the signal of a reference standard solution of known absorbance. Both sample and reference solutions are measured simultaneously by the dual channel LPAS set-up (see Fig. 5). By this method, the influence of long term instability of laser energy and laser beam profile on the photoacoustic signal are eliminated and at the same time the spectrum is converted to absolute absorbance units.

For reference standard solution, a chemically and photochemically stable mixture of aqueous transition metal ions is used. The solution is prepared by mixing 1.0×10^{-3} mol L^{-1} Cr^{3+}, 2.4×10^{-3} mol L^{-1} Co^{2+}, 1.6×10^{-3} mol L^{-1} Cu^{2+} and 7.0×10^{-6} mol L^{-1} $K_2Cr_2O_7$ in H_2O (pH 3.8). Broad absorption bands of each

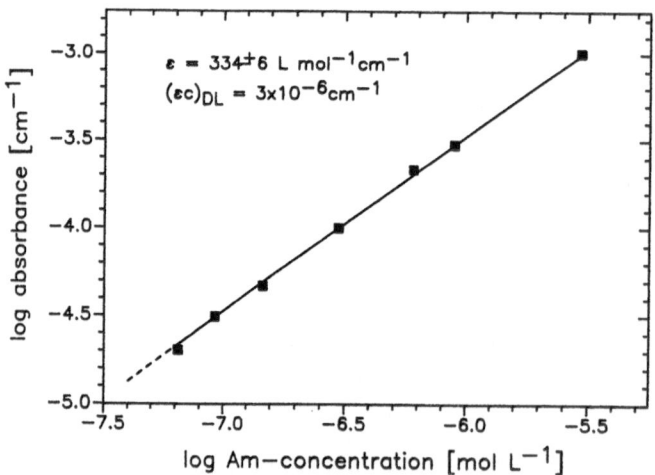

Fig. 17. Absorbance of Am(III) in 1 M Na_2CO_3 (pH = 11.7) at 507.6 nm as a function of the Am concentration determined by LPAS relative to a reference standard. The slope is 1.00 and the derived molar absorptivity is $\varepsilon = 334 \pm 6$ L mol^{-1} cm^{-1} [40]

ion overlap in the range from 350 nm to 800 nm and thus result in a relatively homogeneous flat spectrum. A similar solution was used by Thomson [129] as a reference for absorption spectroscopy. The absorbance of the solution measured over a period of one year is found to remain constant.

A typical example of the spectral work using the reference standard solution [40] is given in Fig. 17, which shows the absorbance calibration of Am(III) in 1 M Na_2CO_3 at pH $= 11.7$ as a function of the Am concentration (from 3×10^{-6} to 6.5×10^{-8} mol L^{-1}). Within this concentration range, the molar absorptivity at the peak maximum at 507.6 nm is calculated to be 334 $\pm$ 6 L mol^{-1} cm^{-1}. The coefficient of variation is 1.8%. The detection limit calculated on three times the background signal is found to be 3×10^{-6} cm^{-1}, which corresponds to 9×10^{-9} mol L^- Am carbonate. When using the reference standard solution cautious attention must be paid to the medium effect of the sample solution on the photoacoustic efficiency.

5 Application to Aqueous Solutions

Since LPAS application to the actinide chemistry is in its infancy, only a limited number of works are available in the published literature. Experiments hitherto performed are confined to either hydrolysis [44], complexation reactions with carbonate [45], EDTA [19] and humate [46] ligands and a variety of speciation works [18–34] for Am(III) [18–28] and to much lesser extent for U(IV), U(VI) [17, 27]; Np(IV), Np(V), Np(VI) [26, 27]; Pu(IV), Pu(VI) [24, 30, 34, 130]. Particular examples of these studies are discussed in this section.

5.1 Speciation Sensitivity

The main advantage of LPAS is a spectroscopic capability for the speciation of aqueous actinide ions in dilute concentrations ($< 10^{-6}$ mol L^{-1}), either their oxidation states or complex forms. Thus the speciation sensitivity is, to a certain extent, conceptually different from the detection sensitivity of an element or isotope. The latter can be better attained by radiometric measurement of actinide nuclides or other conventional analytical methods. The speciation involves the characterization of chemical states of elements concerned. A straightforward speciation can only be made by spectroscopic methods and for this reason the LPAS with its high speciation sensitivity for many aqueous actinide ions is an indispensable instrument. A typical estimation of the speciation sensitivity is given with the Am-EDTA ion [19], because this ion is very stable in aqueous solution. Figure 18 illustrates photoacoustic spectra of the Am-EDTA ion in different concentrations from 10^{-6} mol L^{-1} to 3.5×10^{-8} mol L^{-1}. The characteristic main absorption band of the Am^{3+} ion is the $^7F_0 \rightarrow {}^5L'_6$ transition [42] which appears at 503.2 nm (see Fig. 12). This band is shifted to 506.5 nm upon complexation of Am^{3+} with EDTA and a satellite band at 509 nm becomes distinguishable. The molar absorbance at 506.5 nm determined recently in our laboratory is 601 $\pm$ 3 L mol^{-1} cm^{-1}. With decreasing concen-

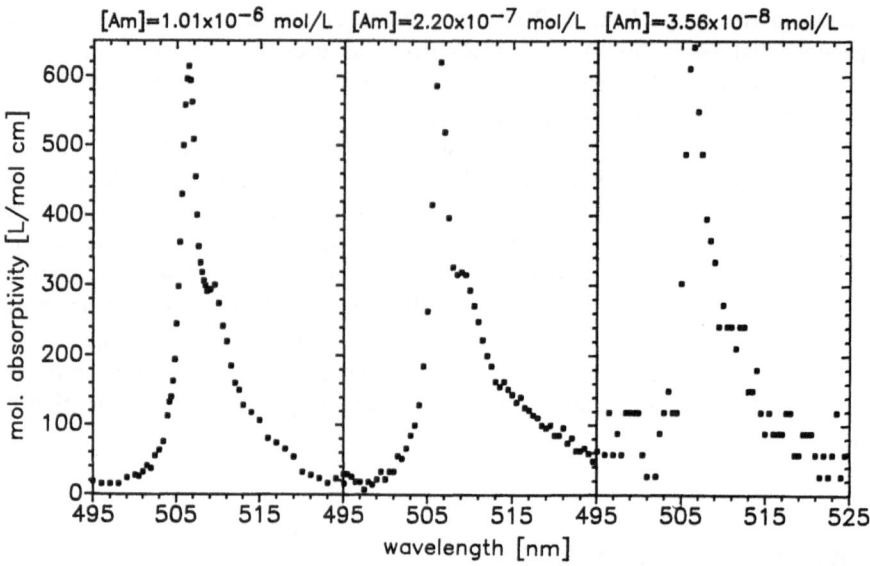

Fig. 18. Photoacoustic spectra of Am(III) in 10^{-3} M EDTA in different Am(III) concentrations [19]

Table 1. The LPAS sensitivities of actinide ions determined in different solutions

Ion	Solution	sensitivity (mol L^{-1})	Ref.
U(IV)	1 M HCl	8×10^{-7} (660 nm)*	[17]
	1 M HClO$_4$	3×10^{-6} (549 nm)	[27]
	1 M NaClO$_4$ (pH 8)	$\sim 10^{-8}$ (539 nm)	[27]
	1 M NaClO$_4$ (pH 11.5)	2×10^{-9} (518 nm)	[27]
U(VI)	0.1 M HClO$_4$	8×10^{-7} (414 nm)	[18, 24]
	1 M HClO$_4$	5×10^{-6} (401.3 nm)	[27]
	1 M NaClO$_4$ (pH 8)	5×10^{-8} (396 nm)	[27]
	1 M NaClO$_4$ (pH 11.5)	5×10^{-9} (393 nm)	[27]
Np(IV)	1 M HClO$_4$	2×10^{-7} (502.7 nm)	[27]
Np(V)	1 M HClO$_4$	2×10^{-6} (617 nm)	[27]
	1 M NaClO$_4$ (pH 8)	1×10^{-6} (617 nm)	[27]
	1 M NaClO$_4$ (pH 11)	4×10^{-8} (605 nm)	[27]
Np(VI)	1 M HClO$_4$	1×10^{-6} (554 nm)	[27]
Pu(III)	1 M HClO$_4$	6×10^{-7} (556 nm)	[27]
Pu(IV)	1 M HClO$_4$	7×10^{-8} (476 nm)	[18, 24]
	1 M HClO$_4$	1×10^{-7} (469 nm)	[27]
	1 M Na$_2$CO$_3$	7×10^{-8} (484 nm)	[19]
	10^{-3} M Gluconic acid (pH 12)	2×10^{-7} (506 nm)	[27]
	10^{-3} M Citric acid (pH 12)	9×10^{-9} (493 nm)	[27]
Pu(VI)	1 M HClO$_4$	3×10^{-8} (831 nm)	[18, 24]
	1 M HClO$_4$	1×10^{-7} (830 nm)	[27]
Am(III)	0.1 M HClO$_4$	7.5×10^{-9} (503.2 nm)	[19]
	1 M HClO$_4$	2×10^{-8} (503.2 nm)	[27]
	10^{-3} M EDTA	5.0×10^{-9} (506.5 nm)	[19]
	1 M Na$_2$CO$_3$ (pH 11.7)	9×10^{-9} (507.6 nm)	[131]
	1 M Na$_2$CO$_3$ (90 °C)	6×10^{-9} (508 nm)	[21]

* Wavelength used for measurement

tration to 3.5×10^{-8} mol L^{-1} the spectroscopic characteristics of Am-EDTA are still well distinguished, although the influence of background scattering becomes larger. Based on the remaining background scattering after the compensation of the main background due to water absorbance, the speciation sensitivity is found to be

$$\alpha = 3 \times 10^{-6} \text{ cm}^{-1} .$$

This value gives rise to the Am-EDTA concentration of 5.0×10^{-9} mol L^{-1}. With the molar absorbance of 334 L mol^{-1} cm^{-1} for $Am(CO_3)_3^{3-}$ and 410 L mol^{-1} cm^{-1} for Am^{3+}, their speciation sensitivities are also determined to be 9.0×10^{-8} mol L^{-1} and 7.5×10^{-9} mol L^{-1} respectively.

The experimental LPAS sensitivities hitherto known for different actinide ions in various solutions are summarized in Table 1. Since the laboratories involved have different LPAS instrumentations and also use different methods for sensitivity determination, the experiments for an actinide solution under the same condition result in somewhat different sensitivities. However, the differences are not dramatically large.

5.2 Hydrolysis Reactions of Am^{3+}

LPAS is used for the speciation of hydrolysed species of Am^{3+} in non-complexing solution ($NaClO_4$) and can thus verify the theoretical speciation based on thermodynamic hydrolysis constants [44]. Such a speciation in dilute concentrations is invaluable for the validation of thermodynamic data determined by other methods. A typical example is shown below for the sensitive region of narrow pH where Am^{3+}, $Am(OH)^{2+}$ and $Am(OH)_2^+$ are co-existant.

The speciation of Am(III) ions is made at first by calculation based on the thermodynamic data; determined by solubility experiments [44], for the relative concentration of each species as a function of pH. The calculated results are then verified for a number of experimental solutions by LPAS. The spectroscopic speciation results are illustrated in Fig. 19. The first spectrum (top) represents the absorption characteristics of the Am^{3+} ion in 0.1 M $HClO_4$, which is shown here as a reference spectrum for comparative purposes. In this acid medium, the Am^{3+} ion is the only species present and has the main absorption band at 503.2 nm accompanied by one large shoulder at 506 nm and another small one at 510 nm. These two shoulder peaks overlap on the increase of pH with absorption bands of Am hydroxide species. The second spectrum taken at pH = 7.13 shows a change in absorption feature, i.e. broadening of the absorption band through decreasing of the main peak at 503.2 nm. This change can be ascribed to the presence of species mixture, namely Am^{3+} and $Am(OH)^{2+}$. For the same pH, the speciation by thermodynamic calculation indicates the presence of the Am^{3+} ion about 75% and the $Am(OH)^{2+}$ ion about 25%. Since the molar absorption coefficient of Am^{3+} is expected to be larger than that of $Am(OH)^{2+}$ and the concentration of the former is greater as well, the latter species may not be easily distinguished in the spectrum. However, in comparison with the spectrum at pH = 1.0, the ratio of the 503.2 nm peak to the 506 nm shoulder is clearly decreased. On increasing pH, as shown in the third spectrum, the main absorption peak at 503.2 nm for Am^{3+} is starting to shift to 504.4 nm which can be assigned to the absorption of $Am(OH)^{2+}$. At the same time the absorption in the region of 510 nm becomes rather pronounced, probably due to the concentration increase of $Am(OH)_2^+$. By changing further to pH = 7.70, the total Am concentration is decreased to 1.78×10^{-6} mol/L and the spectrum (bottom) is changed as well, revealing the presence of $Am(OH)^{2+}$ at 504.5 nm as well as $Am(OH)_2^+$ at 510.5 nm. According to thermodynamic calculation [44], the Am^{3+} ion is expected to be present at pH = 7.70 in small amounts, nevertheless its absorption cannot be well distinguished because this is overlapped by the

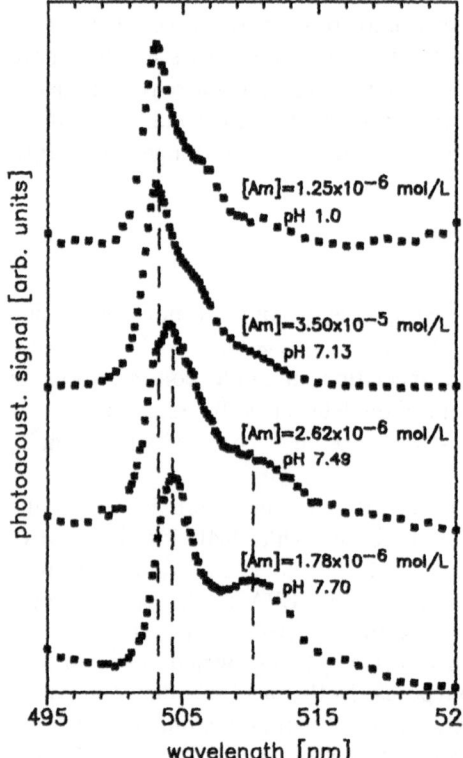

Fig. 19. Photoacoustic spectra of Am(III)-hydrolysis species with increasing pH [44]

$Am(OH)^{2+}$ absorption (504.5 nm). The build-up of $Am(OH)_3$, which starts at pH > 10, cannot be speciated by spectroscopy, since at this pH region the Am concentration is beyond the LPAS speciation limit (10^{-8} mol L^{-1}). However, the solubility data which becomes independent of pH at > 12 indicate clearly the presence of $Am(OH)_3$ (6×10^{-12} mol L^{-1}) in the solution. The spectroscopic speciation has so far verified the species predicted by thermodynamic calculation; the experimental results in Fig. 19 are in fact corroborating thermodynamic calculations for the important region of hydrolysis reactions.

5.3 Carbonate Complexation of Am^{3+}

Another application of LPAS is the speciation of Am-carbonate complexes under deep groundwater conditions [45]. In 0.1 M $NaClO_4$ the carbonate complexation of Am^{3+} is generated under 1 % CO_2 partial pressure in Ar atmosphere by varying pH from 2 to 9.7. With a gradual increase of pH, the Am^{3+} ion originally introduced becomes complexed with the carbonate ion, since the concentration of the latter increases with a power of two as a function of the pH. As the solubility of Am-carbonate $(Am_2(CO_3)_3)$ decreases on increasing pH, the speciation is made by both absorption spectroscopy for higher concentrations ($> 10^{-6}$ mol L^{-1}) and LPAS for lower concentrations ($< 10^{-6}$ mol L^{-1}).

A typical example of the Am-carbonate complexation is shown in Fig. 20, which shows spectra taken by UV-spectroscopy at pH < 7 and by LPAS at pH > 7. At pH $= 2.0$ the Am^{3+} ion is the predominant species showing the absorption maximum at 503.2 nm with double shoulders at 506 nm and 510 nm (cf. Fig. 12). At pH $= 6.4$ a shoulder at 505.5 nm becomes pronounced. Calculation by

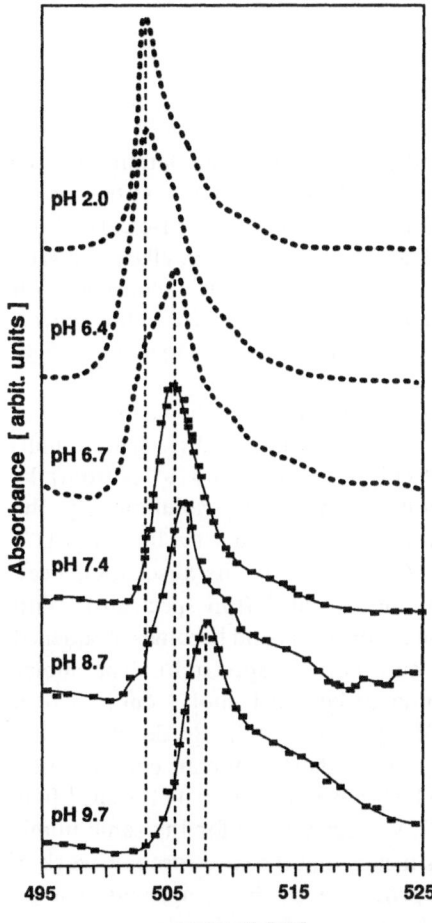

pH 2.0

pH 6.4

Absorbance [arbit. units]

pH 6.7

pH 7.4

pH 8.7

pH 9.7

Fig. 20. Photoacoustic spectra (pH > 7) of Am carbonates together with absorption spectra (pH < 7) in different pH under 1 % CO_2 partial pressure (0.1 M $NaClO_4$). The Am concentration ranges from 5.6×10^{-5} mol L^{-1} to 3.0×10^{-7} mol L^{-1} (from upper to lower spectra) [45]

495   505   515   525

wavelength [nm]

thermodynamic constants determined by solubility experiments [45] indicates that at this pH about 75% Am is the Am^{3+} ion and the rest is a composite of $AmCO_3^+$ and $AmOH^{2+}$ in equal concentrations. At pH = 6.7 the formation of $AmCO_3^+$ becomes greater than $Am(OH)^{2+}$ because the CO_3^{2-} concentration increases more rapidly, by a power of two, rather than the OH^- concentration. The absorption peak at 505.5 nm becomes clearly distinguished at pH = 6.7. On further increasing pH to 7.4, only the absorption band at 505.5 nm is observed. The thermodynamic calculation [45] suggests that the predominant species at this pH should be $AmCO_3^+$ over 80% of the Am concentration and the rest is a mixture of Am^{3+}, $AmOH^{2+}$, $Am(OH)_2^+$ and $Am(CO_3)_2^-$. At pH = 8.7, the absorption peak is further shifted to 506.5 nm which suggests the formation of $Am(CO_3)_2^-$. This is corroborated by thermodynamic calculation indicating that this species is made up of over 95% of Am in the solution and the rest is $AmCO_3^+$. On increasing pH to 9.7, a new absorption peak at 507.8 nm is observed and this absorption band does not change any more with further increasing pH up to 12 [131]. The absorption band at 507.8 nm is ascribed to $Am(CO_3)_3^{3-}$, based on the reasoning that after mono- and di-carbonate formations a natural consequence is the formation of tri-carbonate complex of Am(III).

5.4 Humate Complexation of Am^{3+}

As shown in the above sections (Sects. 5.2 and 5.3) the LPAS speciation with the help of thermodynamic calculation provides a straight forward insight into the chemical reactions of Am^{3+} in dilute aqueous solutions. A similar study is also carried out [46] by LPAS for the Am(III) complexation with aqueous humic acid well characterized previously [132]. This study is conducted with an initial Am^{3+} concentration of 9.1×10^{-7} mol L^{-1} in 0.1 M $NaClO_4$ buffered at pH = 6.0 under Ar atmosphere. This humic acid is a polyelectrolyte with a proton exchange capacity of 5.43 ± 0.16 meq g^{-1} [132].

The LPAS study made for the Am-humate complexation is shown in Fig. 21. The concentration ratio of the functional group in the humic acid (tridentate unit) and the Am^{3+} ion is varied from 0 to 6.3 (Fig. 21: from top to bottom) by adding the humic acid in increasing amount to the Am^{3+} solution. By the dilution effect, the total Am(III) concentration is changed from 9.1×10^{-7} mol L^{-1} to 6.2×10^{-7} mol L^{-1}. The figure shows the Am^{3+} ion with its absorption peak at 503.2 nm and molar absorbance of 410 L mol^{-1} cm^{-1} (top spectrum). With addition of humic acid in increasing amounts the absorption band becomes broadened and the peak maximum is shifted to 506 nm (bottom spectrum). The molar absorbance of the Am-humate complex is found to be 330 L mol^{-1} cm^{-1}, which is the same as the value found for $Am(CO_3)_3^{3-}$ (see Sect. 4.7). By deconvolution of absorption bands for the Am^{3+} ion and Am-humate, the Am-humate complexation constant can be evaluated and is then found to be log β = 6.27 ± 0.04 L mol^{-1} [46]. This value is verified by the experiment with UV-spectroscopy for the same humic acid in the higher Am concentration [133]. This means that the spectroscopic work of LPAS is comparable with other spectroscopic processes for the complexation study and the main advantage is higher sensitivities for many actinide ions.

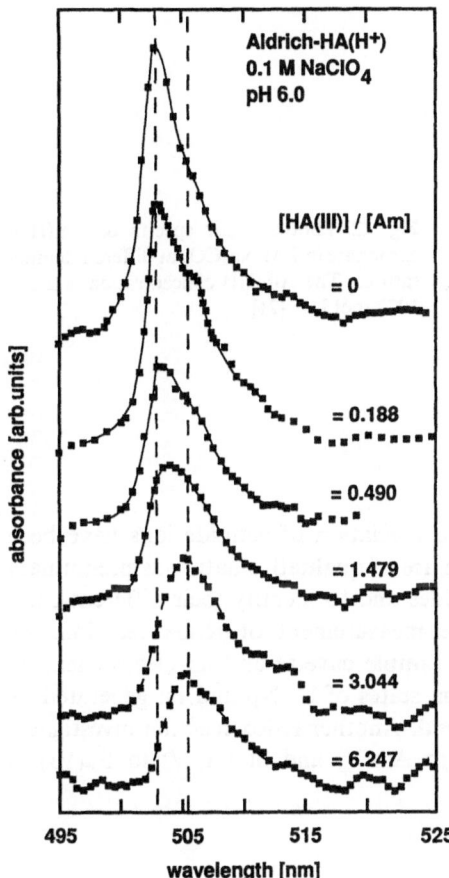

Fig. 21. Photoacoustic spectra of the Am(III)-humate complexation process. The Am(III) concentration varies between 9.1×10^{-7} and 6.2×10^{-7} mol L^{-1}; the humic acid concentration [HA(III)] increases from 0 to 3.9×10^{-6} mol L^{-1} (from top to bottom) [46]

5.5 High Temperature Experiment

The LPAS study of the Am carbonate solution (1 M Na_2CO_3) is conducted at higher temperatures, 30 °C, 60 °C and 90 °C [21] in Argonne National Laboratory. The LPAS spectra are shown in Fig. 22, which shows an interesting change in the photoacoustic signal, increasing with increasing temperature. This phenomenon can be readily explained by Eq. (44) which can be written again in a simplified form for constant experimental conditions:

$$V = \text{const. } \beta v_a^{1/2}/Cp$$

where β is the thermal expansion coefficient, v_a the velocity of sound and Cp the specific heat of the medium. For the temperature region under investigation, the temperature dependent variations of v_a and Cp are minimal [107, 108, 134], while the variation of β is increased about 2.3 times [135] from 30 °C to 90 °C. This means that the thermal expansion of the medium is the main cause of the increase in the photoacoustic pulse height with increasing temperature as observed in Fig. 22.

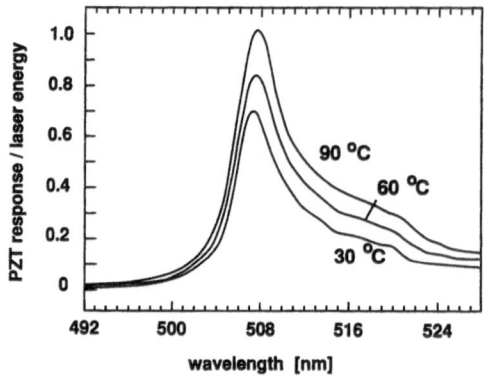

Fig. 22. Photoacoustic spectra of Am(III)-carbonate in 1 M Na_2CO_3 at different temperatures. The Am(III) concentration is 2.5×10^{-6} mol L^{-1} [21]

5.6 Other Investigations

A number of different investigations on the speciation of actinide ions have been performed by LPAS. However most of them are of qualitative nature as preliminary studies to speciate oxidation states of actinides and to identify their different complexes. The LPAS is used for an on-line measurement of redox reactions by connecting a circulation system between a sample cuvette and an electrochemical cell [27]. With this system different oxidation states of U, Np and Pu generated by the electrochemical cell have been investigated. Another redox reaction investigated is an autoradiolytic oxidation of Am(III) to Am(V) and of Pu(IV) to Pu(VI) in

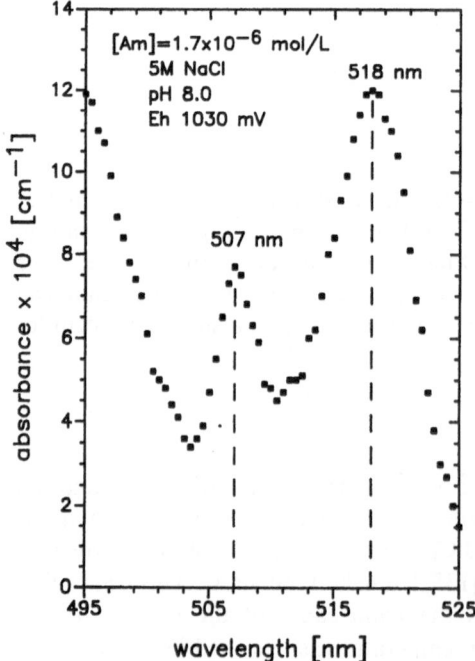

Fig. 23. Photoacoustic spectrum of Am(III) and Am(V) carbonates in 5 M NaCl: autoradiolytic oxidation of Am(III) to Am(V); Am-concentration = 1.7×10^{-6} mol L^{-1}; pH = 8.0 and Eh = 1030 mV; Am(III)carbonate at 507 nm and Am(V)carbonate at 518 nm [33]

concentrated saline solutions [30, 32]. With this reaction the Am(V) carbonate complex as well as the Pu(VI) chloride complex are spectroscopically identified in dilute concentrations [136]. A typical LPAS spectrum of Am(III) and Am(V) carbonate complexes in 5 M NaCl is shown in Fig. 23, in which Am(V) is produced by an autoradiolytic oxidation of Am(III) [33].

6 Application to Natural Aquifer Solutions

Of considerable interest is the LPAS application to the direct speciation of actinides in natural aquifer systems, where the solubility of actinides is in general very low and multicomponent constituent elements as well as compounds are in much higher concentrations than actinide solubilities [28]. The study of the chemical behaviour of actinides in such natural systems requires a selective spectroscopic method of high sensitivity. LPAS is an invaluable method for this purpose but its application to the problem is only beginning. A few examples are shown for the purpose of demonstration.

6.1 Speciation of Am(III) in Groundwater

The Am^{3+} ion in 0.1 m $NaClO_4$ (pH = 6) is introduced to one of the Gorleben groundwaters (Gohy-214) which is kept under inert gas atmosphere (Ar + 1% CO_2). The applied CO_2 partial pressure approximately corresponds to the condition found in Gorleben aquifer systems. The groundwater contains about 5 mg L^{-1} DOC (dissolved organic carbon), of which a substantial fraction is humic and fulvic acid [137]. After 3 months of conditioning, the groundwater is filtered by a Nuclepore filter with a pore size of 400 nm. The LPAS speciation of the filtrate [33] is shown in Fig. 24 (spectrum (a)). Comparing with the previous spectra of Am(III) (Figs. 19–21), the absorption peak observed at 506.0 nm indicates either the carbonate or humate species, while the FWHM value of 4.8 nm suggests the Am carbonate. In consideration of the total carbonate concentration (approx. 3×10^{-4} M) and the humic/fulvic acid concentration (approx. 5×10^{-5} eqmol L^{-1}), and taking into account their complexation constants with Am(III) [44–46], both carbonate and humate (also fulvate) species of Am(III) can be present in this groundwater.

In order to verify the actual chemical state, the solution is transfered from the Ar + 1% CO_2 atmosphere to an Ar atmosphere and the pH is simultaneoulsy lowered to 4.0 in order to decrease the carbonate concentration in the solution. By this process, the Am carbonate species are decomposed to produce the Am^{3+} ion. Upon releasing the partial pressure of CO_2, the colloid generation takes place in the solution which is then filtered at 400 nm pore size. The Am concentration is decreased about 16% to 4.3×10^{-7} mol L^{-1}. The photoacoustic spectrum of the filtrate is shown in Fig. 24 (spectrum (b)), which illustrates the peak shift to 503.2 nm, suggesting the production of the Am^{3+} ion from decomposition of its carbonates (cf. Fig. 20). The high shoulder at 506.0 nm indicates the coexistence of another chemical state of Am(III). Since the carbonate concentration in the solution at pH = 4.0 becomes

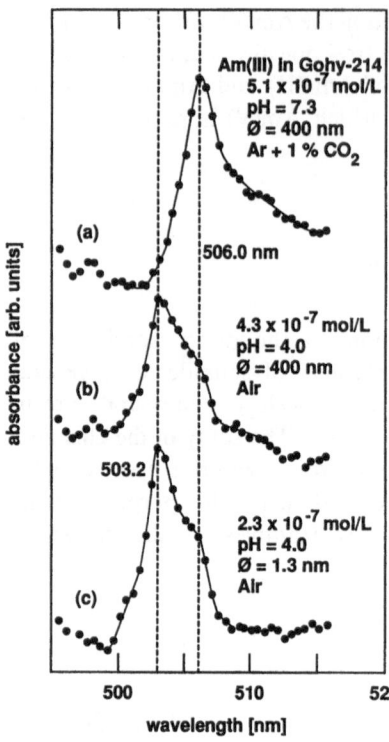

Fig. 24. Speciation of Am(III) by LPAS in one of the Gorleben groundwaters (Gohy-214): (a)-under Ar + 1% CO_2 atmosphere after filtration at 400 nm pore size; (b) — under normal atmosphere (air) after filtration at 400 nm; (c)-under normal atmosphere after filtration at 1.3 nm [33]

negligibly small ($<10^{-7}$ M), the shoulder peak is suspected to indicate the presence of Am humate species. To confirm the presence of Am humates, the solution is further filtered with the smaller pore size of 1.3 nm. The Am concentration in the filtrate is decreased about 47% to 2.3×10^{-7} mol L^{-1} and the spectrum (c) shows the unambiguous identification of the Am^{3+} ion with the better pronounced peak at 503.2 nm. The relative height of the shoulder peak is decreased by filtration at 1.3 nm pore size but the peak does not disappear. This remaining shoulder peak may be ascribed to Am fulvates which are, in general, much smaller in size and remain more stable over a wider range of pH than Am-humate.

From the systematic follow-up of the spectral work with the aid of ultra-filtration, as well as the change in the partial pressure of CO_2 and pH, the original chemical state of Am(III) in the given groundwater is confirmed as being the $Am(CO_3)^+$ ion mixed to some extent with both Am humates and fulvates [33].

6.2 Speciation of Am-pseudocolloids in Groundwater

The presence of actinide pseudocolloids can be verified by LPAS on submitting the solution to some chemical alteration. For example, the Am^{3+} ion traced in the Gorleben groundwaters (Gohy-1011 and Gohy-1012) is speciated in order to elucidate the behaviour of the ion in these groundwaters [24, 29]. The Am traced groundwaters are stored under Ar atmosphere with 1% CO_2 partial pressure for six

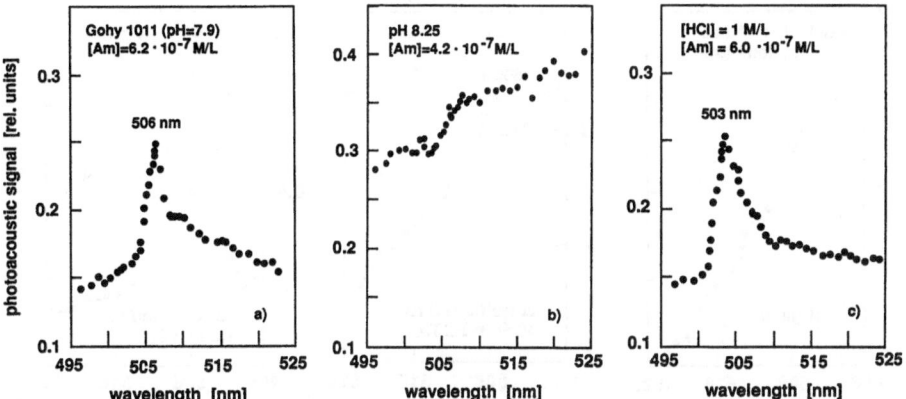

Fig. 25. Speciation of Am(III) by LPAS in one of the Gorleben groundwaters (Gohy-1011) poor in humic acid: **(a)** — Am(III) carbonate; **(b)** — transformation of the carbonate complex to colloids through release of CO_2; **(c)** — recovery of Am^{3+} from colloids by acidification [29]

months. Fig. 25 shows the chemical states of Am^{3+} speciated by LPAS in Gohy-1011 [29]. The LPAS spectrum (a) indicates that the Am carbonate ion ($AmCO_3^+$) is the predominant species under the given conditions, showing its absorption peak at 506 nm [45]. On releasing the CO_2 partial pressure by opening the cuvette and contacting with laboratory atmosphere (CO_2: 0.035%), as shown by spectrum (b), the Am carbonate ion decomposes and Am pseudocolloids are generated. Due to the change in partial pressure of CO_2, the groundwater pH is changed accordingly from 7.9 to 8.25 [28]. Am pseudocolloids are identified by recorded signals of light scattering [24, 29]. The process of changing the CO_2 partial pressure, as shown by spectra (a) and (b), is a situation comparable to normal sampling of deep groundwater. The CO_2 release during the sampling of groundwater will result in groundwater colloids, which will then falsify subsequent laboratory experiments. The Am concentration in the sample of spectrum (b) is decreased to about 32% of the original one (spectrum a)) presumably due to either precipitation or sorption of colloids on the cuvette surface. On acidification to 1 M HCl, the total amount of americium is transformed to the Am^{3+} ion, showing its absorption peak at 503 nm (spectrum (c)). The Am concentration diminished in the sample of spectrum (b) is recovered by acidification to 6.0×10^{-7} mol L^{-1}. A slight difference from the original concentration of 6.2×10^{-7} mol L^{-1} is attributed to dilution on acidification.

The other groundwater (Gohy-1012) containing humic colloids [137] shows different behaviour of Am. LPAS spectra of Am traced Gohy-1012 [29] are given in Fig. 26. The LPAS spectrum (a) without an absorption peak but with a high baseline illustrates the presence of colloids in the solution. The Am carbonate ion is not visible, unlike Am in Gohy-1011, although the CO_2 partial pressure is maintained at 1%. Since the Am concentration remains stable at 2.0×10^{-6} mol L^{-1} for a period of six months, the presence of Am pseudocolloids (humic colloids [137]) appears evident. To a portion of this solution, EDTA is added to make a concentration of 10^{-3} mol L^{-1}. The reason for addition of EDTA is not to destroy humic colloids but to extract only the Am^{3+} ion out of colloids, because the

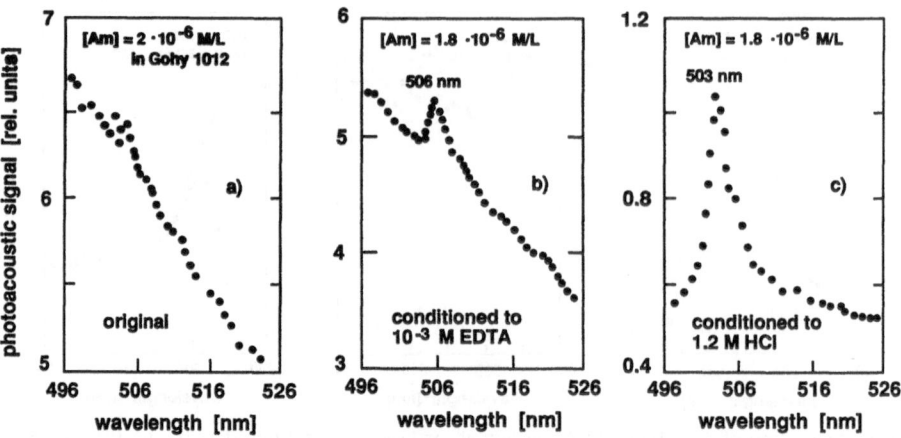

Fig. 26. Speciation of Am(III) by LPAS in one of the Gorleben groundwaters (Gohy-1012) rich in humic acid: (a) — humic colloids of Am(III); (b) — dissolution of Am from humic colloids by EDTA; (c) — precipitation of humic colloids and recovery of Am^{3+} by acidification [29]

complexation power of EDTA with Am^{3+} is much stronger [28] than that of humic acid in the colloids. After six months of conditioning with EDTA, the Am-EDTA complex is observed in the LPAS spectrum (b) with its absorption peak at 506 nm. A steep inclination of the base line with decreasing wave length suggests the presence of humic colloids in the solution. Another portion of the Am traced groundwater is conditioned with 1.2 M HCl. Again after a period of six months, the complete release of the Am^{3+} ion from humic colloids is observed (spectrum (c)) with its absorption peak at 503 nm. In the case of acidification, the humic colloids are precipitated by an excess protonation; the behaviour is comparable with the precipitation of humic acid by acidification [131]. Due to humic colloid precipitation, the spectrum (c) appears devoid of a high base line.

The spectroscopic speciation shown above demonstrates the generation of two different kinds of actinide pseudocolloids in groundwaters with and without humic substances. As evidenced by separate experiments [31], the generation of actinide pseudocolloids is reversible. Pseudocolloids of organic nature are more stable than those of inorganic nature [29]. For the latter, the CO_2 partial pressure plays an important role in their generation.

6.3 Speciation of Am in Saline Groundwater

Since in Germany a migration study of actinides in the overlying aquifer system on salt dome is required, the speciation study has been mostly concentrated on saline groundwaters [19, 24, 30, 34]. One example shown in Fig. 27 is the speciation of Am(III) in a saline groundwater containing about 3 M NaCl. The untreated groundwater containing 8.9×10^{-8} mol L^{-1} Am(III) gives a spectrum with very scattered baseline signals (spectrum (a)), in which an absorption peak at about 504 nm is only to be guessed. After filtration at 30 nm pore size, the filtrate reveals a

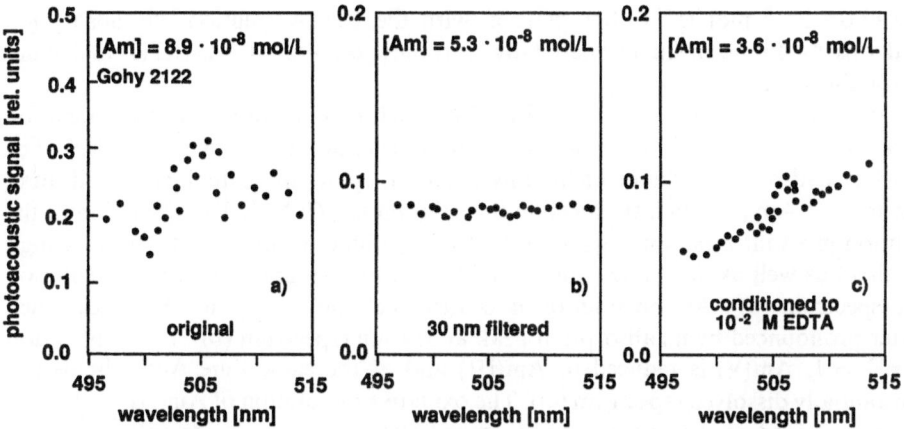

Fig. 27. Specification of Am(III) by LPAS in one of the saline Gorleben groundwaters (3 M NaCl): (a) — unfiltered solution; (b) — filtrate at 30 nm pore size; (c) — the solution conditioned with 10^{-3} M EDTA [19]

stabilization against scattering of signals but no sign of the presence of Am(III) (spectrum (b)), although 5.3×10^{-8} mol L^{-1} Am is in the solution. Since the Am^{3+} ion is suspected to be sorbed in groundwater colloids, EDTA is added to the filtrate to extract Am from the colloids through EDTA complexation. After allowing the solution to stand for over 2 weeks, the Am-EDTA complex can be detected in the spectrum at 506 nm (spectrum (c)). As the Am concentration becomes very

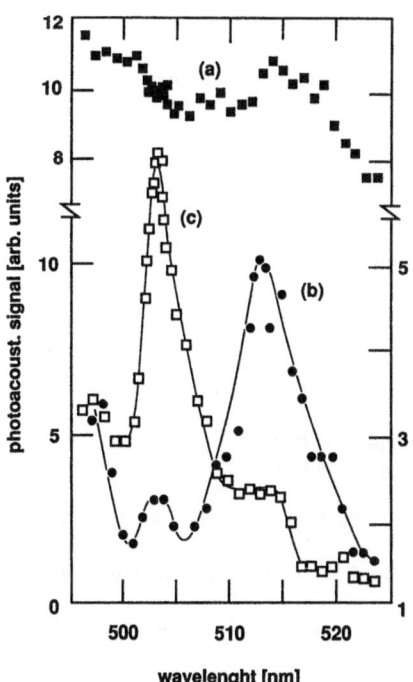

Fig. 28. Speciation of Am(V) by LPAS in one of the saline Gorleben groundwaters (3 M NaCl, pH = 6.7): autoradiolytic oxidation of Am(III) to Am(V): (a) — the unfiltered solution of 7×10^{-6} mol L^{-1} Am showing the presence of AmO$_2^+$ ion (514 nm) and colloids (high base line); (b) — after centrifugation of the solution; (c) — after acidification of the solution to pH = 1 [19]

low $(3.6 \times 10^{-8}$ mol $L^{-1})$ after dilution with the EDTA solution, the absorption peak observed is accordingly small. However, the speciation of Am-EDTA at 506 nm is straightforward.

The second example is shown in Fig. 28, which demonstrates the Am speciation in a solution prepared by dissolving natural rocksalt in groundwater [19]. The Am(III) hydroxide is dissolved in this solution up to the solubility equilibrium. The total ^{241}Am activity introduced in the solution is 2 Ci L^{-1}. The speciation of the solution gives information suggesting that a large amount of colloids (seen by high baseline) as well as the AmO_2^+ ion (seen at 514 nm [32]) are present in the solution (see spectrum (a)). As the solution is centrifuged, the AmO_2^+ ion becomes much better pronounced by its absorption peak at 514 nm (spectrum (b)). By acidification to pH = 1, Am(V) is reduced to Am(III) and at the same time Am colloids are substantially dissolved (spectrum (c)). The oxidative dissolution of Am(III) hydroxide is, as explained above, mainly due to α-radiolysis of the chloride solution which creates a strongly oxidizing medium [30, 32]. The difference in the experiments shown in Figs. 27 and 28 is only due to the radiolysis effect which oxidizes Am(III) to Am(V) when the α-specific activity of ^{241}Am in NaCl solution exceeds to a certain level, i.e. 1 Ci L^{-1} [30, 32].

7 References

1. Bell AG (1880) Am J Sci 20: 305
2. Lüscher E (1984) In: Lüscher E, Coufal HJ, Korpiun P, Tilgner R (eds) Photoacoustic effect, Vieweg, Braunschweig
3. Kreuzer LB (1971) J Appl Phys 42: 2934
4. Patel CKN, Tam AC (1981) Rev Mod Phys 53: 517
5. Tam AC (1983) In: Kliger DS (ed) Ultrasensitive laser spectroscopy, Academic, New York
6. Tam AC (1986) Rev Mod Phys 58: 381
7. Pao HY (1977) Optoacoustic spectroscopy and detection, Academic, New York
8. Somoana RB (1978) Angew Chem Int 17: 238
9. Rosencwaig A (1978) Adv Electron and Electron Phys 46: 207
10. Rosencwaig A (1980) Photoacoustics and photoacoustic spectroscopy, Chem Anal, vol. 57, New York
11. Colles MJ, Geddes NR, Mehdizadeh E (1979) Contemp Phys. 20: 11
12. Kirkbright GF, Castleden SL (1980) Chem Ber 16: 661
13. Lyamshev LM, Sedov LN (1981) Sov Phys Acoust 27: 4
14. Kinney JB, Staley RH (1982) Annu Rev Mater Sci 12: 295
15. West GA, Barrett JJ, Siebert DR, Reddy KV (1983) Rev Sci Instrum 54: 797
16. Zharov VP (1985) In: Letokhov VS (ed) Laser analytical spectrochemistry, Adam Hilger, Bristol
17. Schrepp W, Stumpe R, Kim JI, Walther H (1983) Appl Phys B32: 207
18. Stumpe R, Kim JI, Schrepp W, Walther H (1984) Appl Phys B34: 203
19. Klenze R, Kim JI (1988) Radiochim Acta 44/45: 77
20. Doxtader MM, Maroni VA, Beitz JV, Heavan M (1987) Mat Res Soc Symp Proc 84: 173
21. Beitz JV, Bower DL, Doxtader MM, Maroni VA, Reed DT (1988) Radiochim Acta 44/45: 87
22. Pollard PM, Liezers M, McMillan JW, Philipps G, Thomason HP, Ewart FT (1988) Radiochim Acta 44/45: 95
23. Russo RE, Rojas D, Silva RJ, (in press) Radiochim Acta

24 Stumpe R, Kim JI (1986) Laser-induzierte photoakustische Spektroskopie zum Nachweis des chemischen Verhaltens von Aktiniden in natürlichen aquatischen Systemen, Report RCM 02386 Institut für Radiochemie, Technische Universität München); INIS 17 (23): 81488
25. Doxtader MM, Beitz JV, Reed DT, Bates JK (1988) Speciation of radionuclides in natural groundwaters, Report ANL-88-5 (Argonne National Laboratory)
26. Ewart FT, Liezers M, McMillan JW, Pollard PM, Thomason HP (1988) The development of a laser induced photoacoustic facility for actinide speciation, Report NSS-R103 (UK DOE London)
27. Cross JE, Crossley D, Edwards JW, Ewart FT, Liezers M, McMillan JW, Pollard PM, Turner S (1989) Actinide speciation: further development and application of laser induced photoacoustic spectroscopy and voltametry, Report NSS-R119 (UK DOE London)
28. Kim JI (1986) In: Freeman AJ, Keller C (eds) Handbook on the physics and chemistry of the actinides, vol 4, Elsevier, Amsterdam
29. Buckau G, Stumpe R, Kim JI (1986) J Less-Common Metals 122: 555
30. Büppelmann K, Magirius S, Lierse Ch, Kim JI (1986) J Less-Common Metals 122: 329
31. Kim JI, Buckau G, Klenze R (1987) In: Come B, Chapman (eds) Natural analogues in radioactive waste disposal, Graham and Tratman, London
32. Magirius S, Carnall WT, Kim JI (1985) Radiochim Acta 38: 29
33. Klenze R, Kim JI (1989) Mat Res Soc Symp Proc 127: 985
34. Kim JI, Treiber, Lierse Ch, Offermann P (1985) Mat Res Soc Symp Proc 44: 359
35. Berthoud Th, Mauchien P, Omenetto N, Rossi G (1983) Anal Chim Acta 153: 265
36. Moulin Ch, Delorme N, Berthoud Th, Mauchien P (1988) Radiochim Acta 44/45: 103
37. Grenthe I, Bidoglio G, Omenetto N (1989) Inorg Chem 28: 71
38. Rabinowitch E, Belford RL (1964) Spectroscopy and photochemistry of uranyl compounds, Pergamon, London
39. Beitz JV, Jursich G, Sullivan JC (1989) In: Silber HB, Morss LR, Delong LE (eds) Rare earth 1988, Elsevier Sequoia, Lausanne
40. Klenze R, Kim JI, Wimmer H (in press) Radiochim Acta
41. Decambox P, Mauchien P, Moulin V, Moulin Ch (in press) Radiochim Acta
42. Carnall WT (1986) J Less-Common Metals 122: 1
43. Katz JJ, Seaborg GT, Morss LR (1986) The chemistry of actinide elements, 2nd ed, vols 1 and 2, Chapman and Hall, New York
44. Stadler S, Kim JI (1988) Radiochim Acta 44/45: 39
45. Meinrath G, Kim JI, (in press) Radiochim Acta
46. Kim JI, Buckau G, Bryant E, Klenze R (1988) Radiochim Acta 48: 135
47. Yoshihara K, Hiraga M, Izawa G, Braun T (1988) J Radioanal Nucl Chem Lett 126: 87; (in press) Nucl Instrum Methods
48. Mercadier ME (1881) Phil Mag 11: 78
49. Tyndall J (1881) Proc Roy Soc 31: 307
50. Preece WH (1881) Proc Roy Soc 31: 506
51. Röntgen WC (1881) Ann Phys u Chem 12: 155
52. Lord Rayleigh (1881) Nature 23: 274
53. Veingeror HL (1938) Dokl Akad Nauk USSR 19: 687
54. Luft KF (1943) Z techn Phys 24: 97
55. Kreuzer LB, Patel CKN (1971) Science 173: 45
56. Patel CKN (1978) Science 202: 157
57. Patel CKN, Kerl RJ, Burkhardt EG (1977) Phys Rev Lett 38: 1204
58. Patel CKN (1978) Phys Rev Lett 40: 535
59. Harshbarger WR, Robin HB (1973) Acc Chem Res 6: 329
60. Rosencwaig A (1973) Opt Comm 7: 305
61. Rosencwaig A (1977) Rev Sci Instrum 48: 1133
62. Patel CKN, Tam AC (1979) Appl Phys Lett 34: 467
63. Kohanzadeh Y, Whinnery JR, Carroll MM (1975) J Acoust Soc Am 57: 67
64. Hordvik A (1977) Appl Opt 16: 2827
65. Patel CKN In: Lüscher E, Coufal HJ, Korpiun R, Tilgner R (eds) Photoacoustic effect, Vieweg, Braunschweig
66. Beitz JV, Hessler JP (1980) Nucl Technol 51: 169

67. Tam AC, Patel CKN (1979) Appl Opt 18: 3348
68. Sawada T, Oda S, Shimizu H, Kamada H (1979) Anal Chem 51: 6
69. Mori K, Imasaka T, Ishibashi N (1982) Anal Chem 54: 2034
70. Lahman W, Ludewig HJ, Welling H (1977) Anal Chem 49: 549
71. Berthoud T, Delorme N, Mauchien P (1985) Anal Chem 57: 1216
72. Gordon JP, Leite RCC, Moore RS, Porto SPS, Whinnery JR (1965) J Appl Phys 36: 3
73. Imasaka T, Miyaishi K, Ishibashi N (1980) Anal Chim Acta 115: 407
74. Miyaishi K, Imasaka T, Ishibashi N (1981) Anal Chim Acta 124: 381
75. Askar'yan A, Prkhorou AM, Chanturiga GF, Shipulo GP (1963) Sov Phys JETP 17: 1463
76. Hu CL (1969) J Acoust Soc Am 46: 728
77. Bunkin FV, Kaslov NV, Komissarov VM, Kuzmin GP (1971) JETP Lett 13: 341
78. Botygina NN, Bukatyi VI, Khmelevtsov SS (1976) Sov Phys Acoust 22: 368
79. Cleary SF, Hamrick PE (1969) J Acoustic Soc Am 46: 1037
80. Bell CE, Maccabee BS (1974) Appl Opt 13: 605
81. Sigrist M, Kneubühl F (1976) J Appl Math Phys 27: 517
82. Carome EF, Clark NA, Moeller CE (1964) Appl Phys Lett 4: 95
83. Gournay LS (1966) J Acoust Soc Am 40: 1322
84. White RM (1963) J Appl Phys 34: 3559
85. Lai HM, Young U (1982) J Acoust Soc Am 72: 2000
86. Nelson ET, Patel CKN (1981) Opt Lett 6: 354
87. Sigrist MW, Kneubühl FK (1978) J Acoust Soc Am 64: 1652
88. Atalar A (1980) Appl Optics 19: 3204
89. Naugol'nykh (1977) Sov Phys Acoust 23: 98
90. Lai HM, Suen WM, Young K (1981) Phys Rev Lett 47: 177
91. Lai HM, Suen WM, Young K (1982) Phys Rev A 25: 1755
92. Hunt FV (1972) In: Gray DE, Mc Graw-Hill (eds) American Institute of Physics Handbook, New York
93. Gertsen Ch, Kneser HD, Vogel H (1974) Physik, Springer, Berlin Heidelberg New York New York
94. Mende D, Simon G (1974) Physik-Gleichungen und Tabellen, VEB Fachbuchverlag, Leipzig
95. Landau LD, Lifschitz EM (1959) Fluid Mechanics, Pergamon, New York
96. Farrow MM (1978) Appl Optics 17: 1093
97. Blitz J (1967) Fundamentals of ultrasonics, Plenum, New York
98. White DL (1964) In: Mason WP (eds) Physical acoustics, Academic, New York
99. Von Gutfeld RJ, Melcher RL (1977) Appl Phys Lett 30: 257
100. Fox JA (1974) Appl Optics 13: 1760
101. Peercy PS, Jones ED, Bushnell JC, Gobeli GW (1970) Appl Phys Lett 16: 120
102. Kittel Ch (1976) Einführung in die Festkörperphysik, Oldenbourg, München
103. Breuer HD (1984) In: Lüscher E, Coufal HJ, Korpiun P, Tilgner R (eds) Photoacoustic effect, Vieweg, Braunschweig
104. Meyer FJ (1972) Ph D thesis, Techn. Hochschule Aachen
105. Vernitron (1985) Bulletin 6611/F and 66017/B, Thornhill, Southampton, England
106. Valvo (1981) Piezoxide (PXE), Valvo Unternehmensbereich Bauelemente der Philips GmbH
107. Lovett JR (1969) J Acoust Soc Am 45: 1052
108. Greenspan M, Tschiegg CE (1957) J Research NBS 59C: 249
109. Brau CA, Ewing II (1975) Appl Phys Lett 27: 350
110. Basting D (1979) Laser + Elektro-Optik 4
111. Hohla KL (1982) Laser Focus, June 67
112. Shank CV, Dienes A, Trozzolo AM, Myer JA (1970) Appl Phys Lett 16: 405
113. Gauglitz G (1983) Praktische Spektroskopie, Attempto, Tübingen
114. Demtröder W (1977) Grundlagen und Techniken der Laserspektroskopie, Springer, Berlin Heidelberg New York
115. Bonch-Bruevich AM, Razumova TK, Starobogatov ID (1975) Tech Phys Lett 1: 26
116. Lahmann W, Ludewig HJ (1977) Chem Phys Lett 45: 177
117. Bonch-Bruevich AM, Razumova TK, Starobogatov ID (1977) Opt Spectrosc (USSR) 42: 82
118. Farrow MM, Burnham RK, Auzanneau M, Olsen SL, Purdie N, Eyring EM (1978) Appl Opt 17: 1093

119. Sladky P, Danielius R, Sirutkaitis V, Boudys M (1977) Czech J Phys 827: 1075
120. Burt JA (1979) J Acoust Soc Am 65: 1164
121. Oda S, Sawada T, Kamada H (1978) Anal Chem 50: 865
122. Voigtman E, Jurgensen A, Winefordner J (1981) Anal Chem 53: 1442
123. Thompson WR (1935) Annals of Math Stat 6: 214
124. Sachs L (1971) Statistische Auswerteverfahren, Springer, Berlin Heidelberg New York
125. Bernkopf M (1984) Dissertation, Institut für Radiochemie, Technische Universität München
126. Kim JI, Kanellakopulos B (1989) Radiochim Acta 48: 145
127. Wimmer H, Klenze R, Kim JI (to be published)
128. Querry MR, Cary PA, Waring RC (1978) Appl Opt 17: 3587
129. Thomson LC (1946) Trans Faraday Soc 42: 663
130. Eiswirth M, Kim JI, Lierse Ch (1985) Radiochim Acta 38: 197
131. Wimmer H (1989) Thesis (Diploma), Institut für Radiochemie, Technische Universität München
132. Kim JI, Buckau G, Li GH, Duschner H, Psarros N (in press) Radiochim acta
133. Kim JI, Rhee DS, Buckau G (in press) Radiochim Acta
134. CRC Handbook of Chemistry and Physics (1982) 63rd ed., CRC, Boca Raton
135. Kell GS (1967) J Chem Eng Data 12: 66
136. Kim JI, Lierse Ch, Büppelmann K, Magirius S (1987) Mat Res Soc Symp Proc 84: 603
137. Kim JI, Buckau G, Zhuang W (1987) Mat Res Soc Symp Proc 84: 747

Author Index Volumes 151–157